Editors:
R.W. Johnson, Canberra (Australia)
G.A. Mahood, Stanford (USA)
R. Scarpa, L'Aquila (Italy)

K. Bell   J. Keller   (Eds.)

Carbonatite Volcanism

Oldoinyo Lengai and
the Petrogenesis of Natrocarbonatites

With 65 Figures, Some in Colour, and 38 Tables

 Springer-Verlag
Berlin Heidelberg New York
London Paris Tokyo
Hong Kong Barcelona
Budapest

Professor KEITH BELL
Ottawa-Carleton Geoscience Centre
Dept. of Earth Sciences
Carleton University
Ottawa, Ontario
Canada K1S 5B6

Professor JÖRG KELLER
Institut für Mineralogie, Petrologie und Geochemie
Universität Freiburg
Albertstr. 23b
79104 Freiburg
Germany

This volume is a Contribution of the IGCP Project 314 of IUGS-UNES(
on Alkaline and Carbonatitic Magmatism

Library of Congress Cataloging-in-Publication Data. Carbonatite volcanism: Oldoinyo Lengai ar
genesis of natrocarbonatites/K. Bell, J. Keller (eds.).   p.   cm. – (IAVCEI proceedings in volume
Includes bibliographical references and index.
ISBN-13: 978-3-642-79184-0   e-ISBN-13: 978-3-642-79182-6
DOI: 10.1007/ 978-3-642-79182-6
                    1. Carbonatites – Tanzania – Oldoinyo Lengai Region – Congresses.
ism – Tanzania – Oldoinyo Lengai Region – Congresses. 3. Oldoinyo Lengai (Tanzania) –
I. Bell, Keith, 1938– . II. Keller, J. (Jörg), 1938– . III. Series. QE462.C36C34 1994 :
94-23059

© Springer-Verlag Berlin Heidelberg 1995
Softcover reprint of the hardcover 1st edition 1995

The use of general descriptive names, registered names, trademarks, etc. in this publication do

Dedication

This volume is dedicated to the memory of Katia and Maurice Krafft who died on June 3rd, 1991, in a nuée ardente from Unzen Volcano in Japan. The Kraffts' interest in volcanoes was a consuming passion that was legendary among scientists in the volcanological community.

The successful expedition to Oldoinyo Lengai in June, 1988, was the result of the inspiration, efforts and organizational skills of Katia and Maurice. This expedition led to renewed interest in Oldoinyo Lengai, and to the first film footage of the eruptive activity of carbonatitic magmas showing the exceptional fluidity of the carbonatite flows.

We hope that the dedication of this book to Katia and Maurice Krafft will contribute to the memory of these extraordinary people. Their professionalism was admired and their friendship was cherished by volcanologists the world over.

Preface

During the last few years, carbonatites have received a considerable amount of attention. Some of this interest was no doubt kindled by the importance of volatiles in the Earth's mantle, particularly CO_2, by the fact that carbonatites can be used to monitor the chemical evolution of the sub-continental upper mantle, and by the fact that carbonatites may be effective metasomatizing agents at both mantle and crustal levels.

The interest in Oldoinyo Lengai has extended over at least 100 years, but it was not until the eruptions of 1960, when the unique carbonatitic nature of its lavas was recognized, that the volcano took on special significance in volcanology and igneous petrology. The recognition of carbonatitic flows coincided with the first successful laboratory experiments carried out on carbonatitic melts. Since then, Oldoinyo Lengai has formed a cornerstone in all carbonatite discussions. It is probably true to say that the findings from Oldoinyo Lengai have dominated our ideas about carbonatites, in spite of the fact that the alkali-rich, natrocarbonatitic lavas of Oldoinyo Lengai are markedly different from other carbonatites.

Oldoinyo Lengai lies in the East African rift in northern Tanzania near the Kenya-Tanzania border, south of Lake Natron, and was considered for many years to be fairly inaccessible. It was therefore surprising that in one year, 1988, two expeditions visited Oldoinyo Lengai – the Krafft-Keller expedition in June and the Dawson expedition in November. Findings from these two expeditions stimulated renewed interest in Oldoinyo Lengai, particularly the findings that the temperatures and viscosities of erupting natrocarbonatite flows were among the lowest measured in terrestrial materials. Also, detailed petrographic and geochemical analyses, long awaited on fresh samples, became available.

At the IAVCEI International Volcanological Congress in Mainz in 1990 a symposium devoted to carbonatite volcanism, convened by Bell and Keller, included a session dedicated solely to Oldoinyo Lengai. This meeting provided the stimulus that resulted in this volume. The aim of the volume is to document the most recent findings from Oldoinyo Lengai, and to place the volcano and its products into the context of mantle-generated melts and their evolution. Although we know that this volume will be of interest to most volcanologists, and to all petrologists involved in carbonatite research, we also hope that it will stimulate other earth scientists to take an active role in studying these fascinating rocks.

The editors would like to thank the following for ensuring that all manuscripts reached a high standard: D.K. Bailey, P. Baker, D.S. Barker, T. Bottinga,

G. Brey, M. Condomines, P. Deines, G.N. Eby, S. Foley, J.B. Gill, J. Gittins, D.H. Green, J. Guest, S.E. Haggerty, A.L. Hofmann, D.D. Hogarth, C. Jaupart, A.F. Koster van Groos, K.T. Kyser, A.N. Mariano, B.O. Mysen, H. O'Neill, P.L. Roeder, M.F. Roden, F. Spera, G.R. Tilton, A.H. Treiman, J. Wolff, A.R. Woolley and P.J. Wyllie. We also appreciate the support and patience of Dr. W. Engel of Springer-Verlag. We would finally like to thank Gail Mahood, the series editor for the IAVCEI Proceedings in Volcanology, whose perceptive comments and insights added greatly to the volume.

Ottawa, Canada   KEITH BELL
Freiburg, Germany   JÖRG KELLER
October 1994

Contents

List of Contributors*

Bell, K.   100, 137[1]
Browning, P.   47
Dawson, J.B.   4, 23, 37, 47, 100, 137
Fallick, A.E.   47
Falloon, T.J.   191
Green, D.H.   191
Hamilton, D.L.   163
Hoefs, J.   113
Jackson, D.   47
Keller, J.   4, 70, 87, 113

Kjarsgaard, B.A.   148, 163
Koberski, U.   87
Norton, G.E.   23, 37, 47
Nyamweru, C.   4
Peterson, T.D.   148, 163
Pinkerton, H.   23, 37, 47
Pyle, D.M.   23, 37, 47, 124
Spettel, B.   70
Sweeney, R.J.   191

---

* The addresses of the authors are given on the first page of each contribution.
[1] Page on which contribution begins.

Introduction

As the only known active carbonatite volcano, Oldoinyo Lengai plays an important role in evaluating the genesis and evolution of carbonatites. Since the pioneering work by J.B. Dawson on the natrocarbonatite lavas of the 1960–61 eruption, this volcano has held an understandable fascination for volcanologists, geochemists and petrologists, and has resulted in considerable debate as to whether the natrocarbonatite is a primary or derivative melt.

Oldoinyo Lengai is an unusual volcano on many counts. The uniqueness of its carbonatitic products lies in the fact that they are natrocarbonatites, extremely alkali-rich, with >30 wt% Na_2O. Low ferromagnesian element contents together with extreme incompatible element concentrations indicate a highly fractionated melt. Natrocarbonatites are also enriched in volatiles, yet are essentially anhydrous. Their mineralogy is dominated by phases unknown from elsewhere, or found rarely as high-temperature magmatic minerals. The eruptive temperatures of 500 to 590°C of the carbonatite melts and their viscosities are among the lowest yet measured for terrestrial lavas.

The natrocarbonatite from Oldoinyo Lengai was considered by many to be similar in composition to the parental magmas that generated all or most carbonatites, and with this acceptance went the assumption that most carbonatites crystallized from a melt that was originally rich in Na. The fact that few carbonatites now contain significant quantities of alkalies was attributed to loss of Na and K dissolved in exsolved fenitizing fluids, rich in either H_2O or Cl or F, at low pressures, particularly at crustal levels. This view, however, was certainly not shared by all. The alternative model assumed that most carbonatites were generated from a melt that was not alkali-rich and that natrocarbonatites represent the products of derivative magmas. The high Na_2O and low MgO contents of natrocarbonatites are attributed to differentiation processes such as crystal fractionation or liquid immiscibility.

The findings outlined in this volume cover several of these problems and in doing so show the complexities of magmatic evolution within a single carbonatite-nephelinite eruptive centre. The two expeditions to Oldoinyo Lengai made during 1988, the Krafft-Keller expedition in June and the Dawson expedition in November, brought about renewed interest in this volcano and resulted in comprehensive petrological studies coupled with several new findings.

Some of the first observations of volcanic activity at Oldoinyo Lengai were made during the second half of the 19th century, and the volcano was first climbed by the German geographer Jaeger in 1904. The description of "soda mudflows" from Oldoinyo Lengai by Reck (1914) was probably the first, unwitting description of natrocarbonatite lava, identified as such after the 1960 lava eruptions. A maximum age of 0.32 Ma has been established for initiation of volcanic activity at Oldoinyo Lengai. Phonolites, nephelinites and melilitites, predominantly as pyroclastics, constitute the main volume of the 2200-m cone. Carbonatites form the summit and have resulted dominantly from the historical activity.

Chapters in this volume can be broadly divided into three groups. The first group documents the volcanic history and new volcanological field observations made mainly during the 1988 eruption. Physical properties, particularly viscosity and temperature of flows, were measured in the field, and although they are quite different from those of silicate melts, the morphology, flow lengths and compound flows show many similarities to those of basaltic lavas.

The second group of chapters documents new analytical and experimental data obtained from both carbonatite and silicate lavas of Oldoinyo Lengai. A detailed mineralogical analysis of natrocarbonatites showing nyerereite and gregoryite as the major phases, including their detailed chemical signatures, are now available and the presence of sub-microscopic sylvite, fluorite, apatite and a great variety of minor accessories is now documented in detail. In the case of the silicate lavas, combeite and wollastonite characterize the extremely peralkaline nephelinites. Cathodoluminescence is used to identify and describe most of the unusual phases and their textures. Trace elements from the Oldoinyo Lengai natrocarbonatites, particularly the REE, indicate melts that have been extremely fractionated. The characteristic chemical signature of natrocarbonatites is marked by high concentrations of Sr, Ba, K, Na, Rb, LREE, Cl and F, accompanied by unusually high abundances of Mo, W, As, Sb and Br. The LREE show extreme fractionation compared to other carbonatites, and the normalized La/Sm ratios of greater than 40 are the highest observed in any magmatic rocks.

The isotopic data are consistent with a mantle origin for the parental melt to natrocarbonatite. The negative linear correlation between the $^{143}Nd/^{144}Nd$ and $^{87}Sr/^{86}Sr$ ratios for both carbonate and silicate flows from Oldoinyo Lengai is similar to the East African carbonatite line, and suggests that the lavas are the products of discrete melting events involving the mixing of two end-members similar to HIMU and EMI recognized from oceanic island basalts. Stable isotopes of fresh natrocarbonatite flows and their phenocryst phases show a very restricted range of isotopic ratios that are decidedly mantle in composition. The isotopic study using short-lived disequilibria in the U-Th system provides important constraints on the timing and mechanism of magma genesis. Modelling of the data from the 1988 natrocarbonatite lavas suggests that segregation and subsequent eruption took place over an interval of 20 to 81 years.

The third group of chapters assesses the various models that have been proposed for the genesis and evolution of the Oldoinyo Lengai lavas. Three models have been proposed for the generation of natrocarbonatite, and all focus on the problem of attaining the high abundances of Na that characterize the natrocarbonatite. The most extreme model attributed the Na to interaction between a silicate melt and either saline brines or trona deposits, common in the East African Rift Valley System, particularly in Lake Natron, that lies close to Oldoinyo Lengai. However, most of the $\delta^{18}O$ and $\delta^{13}C$ values of the fresh natrocarbonatites are lower than those of trona, and Th-U disequilibrium data confirm that the natrocarbonatites are neither fused trona nor the products of interaction between magma and a groundwater reservoir of saline brine. The two models remaining involve quite different differentiation mechanisms: crystal fractionation on the one hand and liquid immiscibility on the other. Several chapters stress the close relationship at Oldoinyo Lengai between natrocarbonatites and peralkaline nephelinites.

At least two primary magma types occur at Lengai, one an olivine nephelinite and the other a melilite nephelinite, both capable of generating highly peralkaline magmas by crystal fractionation and natrocarbonatite magma by liquid immiscibility. The conjugate silicate liquid at low pressures and low temperatures corresponds in chemical composition to a peralkaline wollastonite nephelinite. Although melting of an amphibole-carbonate lherzolite at 5 to 27 kb can generate a Na-rich carbonate liquid, a primary magma of similar composition has to undergo complex crystal fractionation before a natrocarbonatite, of the type found at Oldoinyo Lengai, is produced.

From the findings in the volume we can draw the following firm conclusions about the Oldoinyo Lengai natrocarbonatite lavas:

  i. their isotopic signatures (C, O, Nd, Pb and Sr) are distinctly mantle, and
  ii. their extreme chemical compositions reflect a highly differentiated melt.

On the basis of the evidence presented in this volume it would be unreasonable to assume that natrocarbonatites play a pivotal role in understanding the genesis of all carbonatites. Although such carbonatites are intriguing, the findings in this volume suggest that natrocarbonatites reflect an extreme position in a protracted range of melts associated with nephelinite-carbonatite centres. Although carbonatitic liquids can probably be generated in many different ways either as primary products from the mantle, or by extreme differentiation of either silicate or carbonate melts, natrocarbonatites should be considered the result of an extreme process rarely encountered in nature. The evidence from Oldoinyo Lengai based on field relationships, mineralogy, petrography, and chemical studies appears to be consistent with a model that involves immiscible separation of natrocarbonatite from a carbonated peralkaline nephelinite at low pressures and low temperatures.

Historic and Recent Eruptive Activity of Oldoinyo Lengai

J.B. Dawson[1], J. Keller[2], and C. Nyamweru[3]

Abstract

The eruptive history of Oldoinyo Lengai began <0.37 Ma ago, with eruptions of nephelinitic and phonolitic tuffs and agglomerates from the southern, now extinct, crater. Following a period of erosion, black nephelinitic ashes were erupted starting about 1250 a from a new northern crater, which has been the site of subsequent activity; altered natrocarbonatite lava blocks occur in these tuffs. Natrocarbonatite lava was first recorded in the summit area in 1904, though earlier verbal reports of "snow" on the mountain may refer to carbonatite ash.

Due to the remoteness of the volcano, observation of the activity was sporadic before the last 30 years. From a combination of observations since the 1960s and earlier accounts, it appears that, during this century, violent eruptions of mixed carbonate-silicate ash have been preceded by periods of natrocarbonatite lava extrusion lasting about 8–10 years, during which the active northern crater has been filled. The ash eruptions have resulted in changes in crater morphology, and have been followed by periods of quiescence, prior to renewed lava effusion. Extrusion of peralkaline combeite nephelinite earlier this century (in 1917?), together with evidence for molten silicate material during the 1966 eruption, points to the coexistence of silicate and carbonate magmas at the volcano.

1 Introduction

Oldoinyo Lengai is the youngest of the Neogene-Quaternary volcanoes in northern Tanzania (Fig. 1). It is one of a group of nephelinite-phonolite-carbonatite volcanoes that erupted subsequent to the last major phase of faulting at 1.2 Ma in the Tanzania sector of the Gregory Rift Valley (Dawson 1992). A maximum age for the onset of the activity is given by a date of 0.37 Ma obtained on mica-rich tuffs that are overlain by Oldoinyo Lengai tuffs in an area south-east of the volcano (MacIntyre et al. 1974). The older of the two main volcanic units at Oldoinyo Lengai, the Unit I yellow tuffs and agglomerates which erupted from the extinct southern crater (Dawson 1962a), has been correlated with the Ndutu and Naisiusiu Beds of the Olduvai Gorge succession

[1] Grant Institute of Geology, University of Edinburgh, West Mains Road, Edinburgh EH9 3JW, UK
[2] Institut für Mineralogie, Petrologie und Geochemie, Universität Freiburg, Albertstr. 23b, 79104 Freiburg, Germany
[3] Department of Anthropology, St Lawrence University, Canton, NY 13617, USA

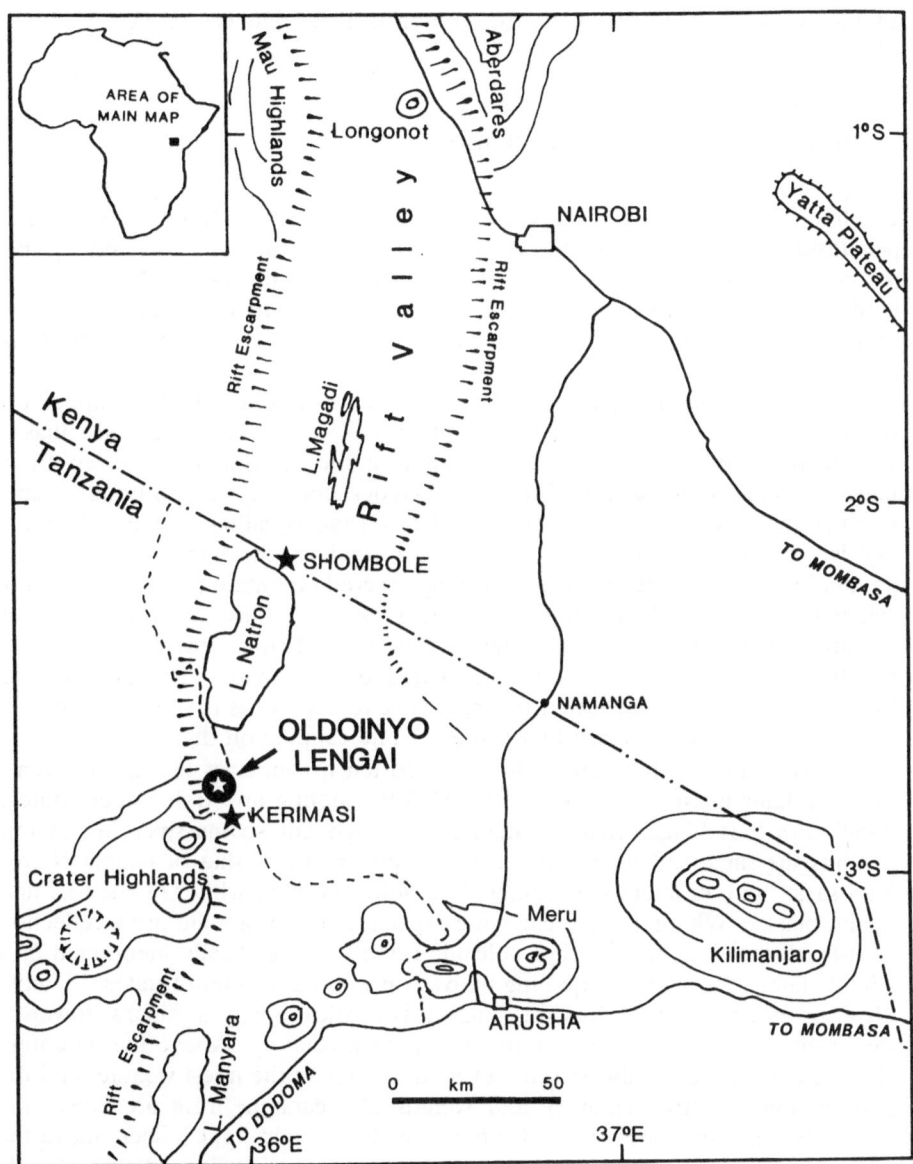

Fig. 1. Location map for Oldoinyo Lengai

on the edge of the Serengeti Plains to the west of the volcano; these beds range in age from about 0.15 to 0.4 Ma (Hay 1976). The younger Unit III black tuffs and agglomerates, erupted from the active northern crater, correlate with the Namorod Ash of the Olduvai succession, dated at about 1250 a. The earliest evidence for natrocarbonatite lava is altered blocks of natrocarbonatite in the Unit III tuffs (Dawson 1993). Altered natrocarbonatite blocks are also found in another younger mappable unit, the Footprint Tuff, that has an age of about 600 a (Hay 1989).

2 First Accounts of the Volcano

The first verbal accounts of the volcano were gleaned from Arab traders whose penetration of the interior of East and Central Africa was motivated by the search for trade, slaves and ivory. Partly because of the warlike Maasai, who limited Arab access to the interior of Kenya and northern Tanzania, knowledge of this part of the continent was extremely limited until the latter half of the 19th century.

Oldoinyo Lengai is first shown on a map dated 14th March 1855, compiled by two missionaries, Erhardt and Rebmann, based at Kisaludini near Mombasa. This map, compiled "in true accordance with the information received from natives", is in the possession of the Royal Geographical Society in London, and is catalogued "Tanzania General 16". The volcano is called "Edonyo Engai", and is annotated "Rain Mtn", "God's Mtn" and "Snow Mtn". The first two annotations are straightforward, indicating a sacred mountain (the word Engai is synonymous for God and rain in Maasai), but the third is intriguing. The elevation of Oldoinyo Lengai is insufficient to sustain a snow cover (unlike the two other "snow mountains" indicated on the map – Kilimanjaro and "Kignea"), so it is possible that this reference to "snow" is the first incognizant record of the carbonatite ash that forms a white capping on the volcano.

The volcano certainly figures in early written accounts of the interior compiled by later missionaries. Wakefield (1870), quoting an Arab trader, states: "Sadi says it is bigger than Kilimanjaro, though not so massive. Its summit exhibits the same radiating and coruscant appearance as that of Kilima Njaro. Sadi says: "one moment it is yellow, like gold; the next white, like silver; and again, black". Whether these changing aspects of the mountain are a reflection of ash fall-out is debatable, but volcanic activity is specifically mentioned in a slightly later report. Again quoting native sources, Farler (1882) states: "Donyo Ngai, or mountain of God, is very high; it is a volcano with a minara (tower or peak) on the top from which smoke is always ascending. Above the mountain there is a black cloud, always seen, even at midday in the finest weather and the sun shining brightly. Thunder and sounds like cannon firing are constantly heard. No fire runs down the sides but at night a bright light is seen above the mountain". Thus, before the onset of European exploration, it was already established that there existed in the remote heart of Maasailand an active volcano sacred to the Maasai – Oldoinyo Lengai, "the Mountain of God".

3 Historic Activity

At the outset, it is worth noting that, in view of the very few ascents of the volcano up to the 1960s, the observation record is probably incomplete and biased in favour of the ash ejection that can be seen from a distance; lava extrusion in the crater may well be underestimated. All the historic eruptions have taken place from the northern crater.

The first European to see the mountain, in 1883, was the German geographer G.A. Fischer, who observed "smoke" rising from the summit and recorded reports by local people of "stretches of fire" on the flanks of the volcano and rumbling noises within the mountain (Fischer 1885). The first attempt to climb the volcano by the biologist O. Neumann in 1894 was thwarted by steep and brittle slopes 150 m below the summit (Wichmann 1894).

4 First Report of Carbonate Lavas

The first scientific account of the volcano and its summit area followed from the ascent of the volcano by F. Jaeger in 1904, during the expedition sponsored by the Otto Winter Foundation; Jaeger's two companions on the climb, Uhlig and Gunzert, stopped 250 m below the summit. In his reports of the expedition, Uhlig (1905, 1907) describes the mountain as composed mainly of tuffs, with some older lavas at the base. He reported the mountain as having an extinct southern crater and an active northern crater with several small cones and fissures that emitted "mudflows" and fumarolic gases (Fig. 2a). With the benefit of hindsight, the description of the "mudflows", which were "covered with efflorescences of a white sodium salt" (Uhlig 1905), is the first, unwitting report of natrocarbonatite lavas. Uhlig also recorded the presence of thick vegetation on the lower slopes, so thick that it could only be penetrated along game trails. This in itself strongly suggests that there had been no major ash eruptions for a considerable time.

On a subsequent visit to the summit with F.Th. Müller on 4 August 1910, Uhlig observed that the northern crater had only a horse-shoe-shaped southern rim immediately below the summit, and lacked a crater rim to the north, west and east. The crater was more like a platform on which there was a central cone from which gas was being emitted. There were also other fissures and vents and "at some openings a black, very hardened mud has appeared". A large 40- to 50-m-high "mud" pinnacle, the Devil's Needle (Teufelsnadel) (Fig. 2b) had formed on the northern rim of the crater since the 1904 visit, and Uhlig noted flows of "soda-mud" on the upper slopes of the volcano (Uhlig and Jaeger 1942). Uhlig's observations clearly are of natrocarbonatite extrusions (lava pinnacles similar to the Devil's Needle were formed on the crater floor in 1988) and, furthermore, indicate considerable activity between 1904 and 1910.

Further ascents of the volcano in 1913 by Reck, and in 1915 by Schulze, resulted in further detailed observations and a photographic record of the summit area and the active crater (Reck 1914, 1924/1925; Reck and Schulze 1921). One of Reck's photographs shows the crater to be very similar to Fig. 2b, and Reck states that "soda-mud" flows from the Devil's Needle and an adjacent cone had flowed down the northern slopes. Schulze's observations show that there had been new lava flows in the crater between December 1913 (Reck's

Fig. 2a,b. The northern crater as seen from the south on 4 September 1904 by Jaeger (**a**), and on the 4 August 1910 by Uhlig and Jaeger (**b**). The prominent Devil's Needle standing at the northern edge of the lava platform (**b**) evidently formed between the taking of these two photos in 1904 and 1910, testifying to copious lava extrusion between those two dates. The two other cones are the NW cone on the rim and the nearer central cone. (Uhlig and Jaeger 1942; Reck 1924/1925)

visit) and Schulze's visit in April 1915. Schulze also records a flow on the western slopes between 1800 and 2000 m.

5 1917 and 1926 Ash Eruptions

The volcano erupted violently in January 1917, causing the local Maasai to sacrifice goats and milk at the foot of the mountain to pacify God (Hobley 1918). It continued in eruption until around June, and ash was distributed "to a distance of twenty-five to thirty miles". Hobley also reports that "lava has flowed for a long distance down the valleys that score the flanks of the mountain, and, in cooling, it has cracked into irregular masses, having the appearance of cakes of grey cement". Hobley's report was second-hand, being based on observations by a Major E.D. Browne, and the implication of lava extrusion during explosive activity gives grounds for caution. Nonetheless, the description of the lava is very reminiscent of more recent lavas, some of which are highly mobile. It may be that Hobley's report refers to lava flows that *preceded* the 1917 eruption, perhaps representing a continuation of lava overspill from vents in the summit area similar to that recorded by Schulze in 1915. From Schulze's report it is clear that the crater lacked walls that might have restricted overspill of further lava flows to the north, west and east.

The 1917 eruption terminated a period of several years of lava extrusion beginning at least in 1904. Ash fall killed off the formerly luxuriant vegetation that clothed the lower slopes of the mountain; it has not recovered fully to the present day.

Altered natrocarbonatite lava, possibly originating from the 1904–1917 effusions, is overlain by nephelinitic tuffs and agglomerates in down-faulted blocks on the south side of the present-day crater (Dawson et al. 1987). Furthermore, at some point before the deepening of the crater by the 1917 eruption, there was minor extrusion of combeite nephelinite lava onto the upper western slopes of the volcano; its stratigraphic position suggests that it post-dates the altered natrocarbonatite in the down-faulted blocks in the southern crater (Keller and Krafft 1990). This is an example of silicate lava extrusion in this century, and there is evidence for a silicate magma chamber beneath the volcano at the time of the mixed silicate/carbonate ash eruption in 1966 (Dawson et al. 1992).

The volcano was next visited in 1921 by Alexander Barns, who observed the mountain from the valley between it and Kerimasi to the south (Barns 1921, 1923). He states "a thin film of vapour arises over the sharply cut edge of the narrow vent but no glow is visible at night". He was much impressed with the ash cover from the 1917 eruption, and observed that "the saline mud shot out by the eruption of Oldoinyo Lengai has erected a low dam connecting it with the Gelei volcano" (Barns 1923). Whether this refers to lava flows is debatable, but as a feature it obviously made an impression on Barns, who was in most respects a keen and reliable observer. The next eruption apparently took place in 1926 and was observed by a Mr. W.H. Billington, who was based at the soda works at

Magadi, 95 km NNE of Oldoinyo Lengai, from which Oldoinyo Lengai is visible in good weather; his observations, recorded by Richard (1942), were of an ash eruption that obscured the south-west horizon for several weeks, and that lasted longer than the 1940 eruption.

The earliest aerial photograph of the crater was taken in January 1930 by Walter Mittelholzer, the flight pioneer and pioneer of aerial photography; it forms part of a series of photographs taken during a flight from Europe to Kilimanjaro (Mittelholzer 1930). This shows a deep crater in the area to the north of the summit (formerly occupied by the flat lava platform), with an even deeper pit in the northern part of the crater. There is no sign of the Devil's Needle, formerly situated on the northern edge of the crater. These major changes most probably resulted from the 1917 eruption, though some further modification could have been due to the poorly documented 1926 eruption.

6 1940 Ash Eruption

The onset of the 1940 ash eruption was first noticed sometime towards the end of July by local Maasai, who were again sufficiently alarmed to sacrifice goats and milk at the foot of the mountain (Richard 1942). The initial activity was characterized by minor puffs of ash but at about 10 am on 31 July, a major initial ash outburst, observed by Mr. Page Jones, a District Officer, quickly developed into a Plinian-type ash pillar which continued for some 7 h. Billington also noted the eruption from Magadi, and reported that it initially was active for a period of 2 to 3 weeks, followed by a pause of several days, after which it began to erupt again. Over this period the ash cloud varied in both height and intensity. An overflight of the volcano on 20th November by Sir Mark Young, the Governor-General of Tanganyika Territory, and Mr. Page Jones, revealed very minor emission of "smoke" from a deep funnel-like hole in the northern crater. Richard (1942) observed Lengai from Magadi on 10th December and it was apparently quiescent, though "looking as if covered with snow". During the eruption, ash had been widely distributed around the volcano, but particularly to the west, where it had reached at least as far as the Loliondo track 100 km distant.

The value of Richard's compilation of the accounts, together with his observations on the ash stratigraphy in the summit area of the volcano in January 1941, was to identify three major phases: (1) a preliminary stage, characterized by small explosions that discharged old material from the vent; (2) the second phase, during which great quantities of gas were released, with discharges sometimes assuming the form of a huge column, preceded by violent explosions that ejected blocks and bombs; and (3) the third or ash phase which did not eject material to such great heights, the ejecta consisting mainly of ash.

Richard drew an analogy between these three eruptive stages and similar phases observed during eruptions of Vesuvius and some of the volcanoes in the East Indies. Richard (1942) also provides a sketch of the profile across the summit area (Fig. 3), the first since the one showing the summit in 1913–1915

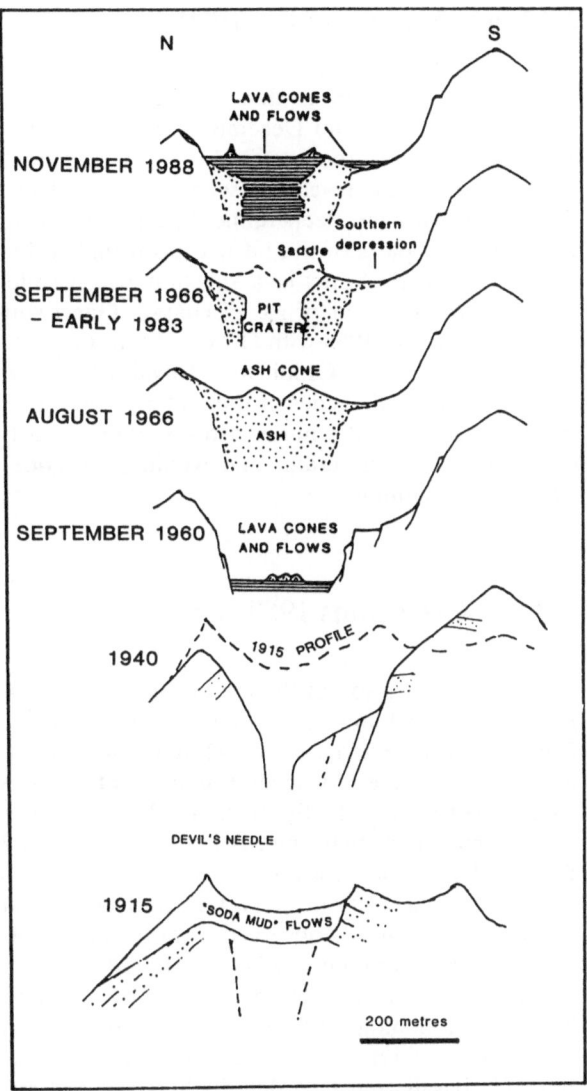

Fig. 3. Profiles across the northern crater. Profile sources are: *1915* Reck and Schulze (1921); *1940* Richard (1942); *1960* Dawson (1962a); *others* Dawson et al. (1990)

(Reck and Schulze 1921), which shows the deepening of the active crater, presumably as a result of the 1917 eruption. On this profile, Richard also shows a profile dated 1926, but since his knowledge of the 1926 eruption appears to have been based on the verbal reports of Billington (who, as far as is known, did not visit the volcano), the 1926 profile must be regarded as speculative.

7 Minor Activity in the 1950s

The next observation is that of Guest (1956), who observed the crater in August, September and December 1954 and in January 1955. On the first visit, lava was splashing from a vent in one of two small cones on a flat crater floor and, at night, the lava was seen to glow. The flatness of the crater floor, contrasting with the deep funnel-like pit shown on Richard's profile, suggests that the lower parts of the pit had been infilled by lava since 1941. On the two later visits in 1954, no lava ejection was visible though there was constant rumbling and the roar of gas discharge. Guest subsequently visited the summit on 24–26 January 1955 with O.M.B. Milton who "had observed and reported an ejection of ash from Oldoinyo L'Engai a few days previously"; some new ash was observed on the sides of the crater cones. A point about this small eruption is that, except for the minor ash discharge seen by Milton, the lava discharge was confined to the crater and would have gone unrecorded without Guest's climb to the summit.

8 Lava Extrusions 1958–66

An aerial photograph of the crater taken on 6 February 1958 by the Air Survey Division of the Lands and Surveys Department, Tanganyika Government (see Dawson 1962a, Fig. 1), shows whitened lava flows on the crater floor, and one major cone, pierced by several minor vents from which small amounts of black spatter had been recently ejected. The latter are firm evidence for a shallow body of magma at that time. A later air photograph of the crater taken by B.E. Pearce-Fleming on 15 March 1960 shows numerous flows of both pahoehoe and aa lavas from a vent on the eastern part of the crater floor.

Lava extrusion was observed from the crater rim in June 1960 (Fig. 4) (C.M. Bristow, pers. commun. to Dawson), and the lavas were sampled following the first-ever descent into the crater made by J.B. Dawson and R. Pickering of the Tanganyika Geological Survey in October 1960. Dawson (1962a) observed that, despite their unique carbonate composition (Dawson 1962b), the lavas behaved like fluid basalt, and recorded discharge of both aa and pahoehoe flows onto the 250-m-diameter crater floor in the September-October 1960 period. He also observed the dramatic colour change from the jet black of newly discharged lava to whitening within 24 h of extrusion; this has proved helpful when interpreting eruption dates from photographs. Subsequently, lava discharge took place till late July 1966; this may have been sporadic, but lava discharge was observed by J.B. Dawson on 7 August, 1961, by Du Bois et al. (1963) in August 1961 (exact date not specified), and on February 2nd 1963 (C. Milton, see Williams et al. 1986), in September 1964 (aerial photograph taken by Y. Malcolm-Coe), and during a visit by the Mountain Club of Kenya on 31 July 1966 (L.A.J. Williams, pers. commun. to Dawson). Over this whole period between August 1960 and July 1966, the activity was characterized by the effusion of small-volume lava

Fig. 4. The active crater, looking north from the summit, June 1960. New black spatter surrounds minor vents, and older whitened flows cover the crater floor. The height of the northern crater wall is approximately 150 m. (Photo C.M. Bristow)

flows and spatter onto the floor of the crater from a succession of small transient cones, vents and lava pools.

9 The 1966–67 Ash Eruption

On 9th August 1966, an airline pilot of East African Airways reported activity at Oldoinyo Lengai in the form of an ash plume up to a level of about 4000 m. Subsequently, pilots operating on the airline route between Johannesburg and Nairobi, which passes almost directly over the volcano, reported ash clouds at flight levels of 10 700 m. The activity in late August has been described and photographed by Dawson et al. (1968). On 20 August, a light ash cloud, rising to no more than 2000 m above the volcano, was drifting northwards and depositing ash on the northern slopes. On 21 August sporadic ash discharge was taking place from a central vent in a new ash cone that had largely filled the previously deep crater (Fig. 5); ash collected on the eastern rim of the crater subsequently proved to be a mixture of carbonate and silicate material, including the rare mineral combeite (Dawson et al. 1989, 1992). A violent eruption began at 14.45 h, and the initial turbulent ash cloud soon developed into a narrow eruption column that rose to at least 3500 m above the cone (Fig. 6); this

Fig. 5. Aerial view from the north-west of the new ash cone that was formed in the crater in August 1966. Compare with the 1966 profile in Fig. 3. (Photo J.B. Dawson)

Fig. 6. The eruption column on 20 August, 1966. (Photo J.B. Dawson)

continued for the rest of the day. The eruption was accompanied by a strong uphill wind and much static electricity; despite violent thunder and lightning within the ash cloud, there was no rainfall. On the following 2 days, activity had decreased to a steady ash plume rising to no more than 500 m above the volcano.

This series of direct observations confirmed the style of eruptive activity inferred by Richard (1942) from his compilation of verbal accounts of the 1940 eruption. Another major ash eruption took place on 3–4 September and a photograph taken on 17th September by Y. Malcolm-Coe shows that the ash cone had partly collapsed to form a pit crater from which a light amount of ash was being emitted. Another major ash eruption took place on 3 October (L.A.J. Williams, pers. comm. to Dawson). During a high-level overflight of the volcano on 27 October, Dawson observed a light ash cloud drifting away to the north-west, and the volcano was completely mantled by white ash. During this series of eruptions ash fell all around the volcano for distances of some 10–20 km but mainly to the north and west; fall out was reported at Seronera (130 km west), at Loliondo (70 km north-west) and Shombole (70 km north). The volcano was quiescent during on overflight on 1 January 1967 by A. Davies. Another major explosive eruption took place on 8–9 July 1967 (Goriatschev 1968). Ash fell at Arusha aerodrome, 110 km to the south east and Wilson Aerodrome, Nairobi, 190 km to the north east, and at both airfields the corrosive dust damaged aluminium aircraft (H. Cameron, pers. comm. to Dawson).

The volcano was climbed in August 1968 by members of a Brathay/Kenya schools expedition. Photographs indicate that the pit crater was deeper than in January 1967, possibly as a result of the July 1967 eruption. The pit occupied the northern part of the main crater, being separated by a narrow ash ridge (later to be called "the Saddle") from a shallow ash-filled depression (later termed "the Southern Depression") overlying the former collapse terraces on the southern side of the main crater. The party saw no ash ejection from the vent but heard rumbling at depth. A party of geologists from the USSR Academy of Sciences, who visited the volcano at the same time, set up a temporary seismic station, and recorded an average of 25 tremors a day over a 4-day period (W.T. Jenkins, pers. comm. to Dawson).

10 Period of Inactivity

The volcano then entered a period of quiescence, though fumarole activity continued. An aerial photo of the crater by W.I. Rose in September 1968 shows no indication of activity (Fig. 7) (also illustrated in Keller and Kraff 1990). Photographs taken looking directly into the pit during overflights of the volcano by Dawson on 6 November 1969 and Krafft in January 1977 also show no activity; comparisons of these photographs show the crater to be virtually unchanged over this period, save for slumping of debris from the east wall of the pit. The crater was unchanged, and there was no visible sign of activity when it was visited by Dawson and M.S. Garson on 1 October 1981; however, there was

Fig. 7. Aerial view showing the inactive pit crater, September 1968. Note the prominent transverse ridge (the Saddle) separating the pit from the Southern Depression. (Photo W.I. Rose)

the sound of sporadic rumbling at depth. Sketches showing the crater during this period, and others contrasting it with the crater both before and after this period of inactivity, are given in Nyamweru (1990). For a profile of the crater between September 1966 and early 1983 see Fig. 2.

11 Resumed Activity 1983

Activity resumed on 1st January 1983 with light ash eruptions. Further ash eruptions, some of considerable size, continued through February and March (summarized by Nyamweru 1988). Ash ejection on the 15 or 16 March is the last record of activity visible from beyond the summit area. From April 1983 onwards, there are relatively frequent (though irregular) records of the appearance of the crater from people who overflew the crater or climbed the mountain. Most of these records are summarized in the Bulletin of the Global Volcanism Network (produced by the Smithsonian Institution, Washington DC; before 1990 it was titled the Seismic Alert Network Bulletin). Lava extrusion in the bottom of the pit crater began in early April 1983 (H. Kroze and J. Fanshawe, pers. commun. to Nyamweru). Considerable volumes of lava were extruded in 1984. A sequence of photographs for this year shows significant filling of the pit; a photograph taken in May 1984 (Fig. 8) shows that the pit had not entirely

Fig. 8. View of the pit crater from the northern crater rim, June 1984. The pit is almost filled with lava flows and spatter cones, but the upper parts of the steep walls of the pit are still visible. (Photo F. Trott)

Fig. 9. The crater from the summit 25 June 1988. The pit crater has completely filled and the whole level of the crater floor had been raised by lava outflows from small cones (some visible on the photo) and fissures. Note also the small cones on the insides of the crater walls and the cone on the northern rim, approximately in the same position as the Devil's Needle in 1910–1917 (see Fig. 2b). The maximum height of the northern crater wall is about 50 m; compare with Fig. 4. (Photo J. Keller)

filled, but another taken in October shows the pit to be completely infilled with lava flows abutting onto the northern slopes of the Saddle. Most of the extrusions took place from vents on the crater floor but between July 1984 and July 1985, a small cone formed on the northern rim of the crater, and several small cones, some extruding small flows, formed on the inner walls of the crater (Fig. 9).

Flows from numerous small cones and lava pools continued to fill up the shallower parts of the crater above the pit and sometime between 26 July and 20 November 1988 began to spill over the Saddle into the Southern Depression (Fig. 10). Continued overspill, with thermal erosion at the base of the rapidly flowing lavas, was documented in November 1988 (Dawson et al. 1990; Nyamweru 1990). Detailed studies of the flow morphology and physico-chemical properties of the lava were made in June 1988 (Krafft and Keller 1989; Keller and Krafft 1990) and November 1988 (Dawson et al. 1990; Pinkerton et al., Chap. 2, this Vol.). Mineralogically and chemically, the lavas are very similar to those extruded in 1960, and comparison of photographs of the crater in the 1960s with those in the 1980s shows the scale and morphology of cones and lava features to be virtually identical. Since late 1988, continued lava extrusion from cones and lava pools in the active sector of the crater has raised the floor of the active part of crater. Lava overspill continued into the Southern Depression,

Fig. 10. The crater from the summit, 23 November 1988. Compare with Fig. 9. The level of the crater floor has again risen due to continued lava extrusion, and flows from the prominent breached cone are spilling into the Southern Depression in the foreground (see 1988 profile, Fig. 3). On the far side of the crater floor is a 15-m-high lava pinnacle similar to the Devil's Needle seen on the northern rim of the crater prior to the 1917 ash eruption. The *scale* is given by the geologist standing to the left of the breached cone. (Photo J.B. Dawson)

Fig. 11. Panoramic view of the crater from the south, 30 September 1992. Compare with Figs. 9 and 10. The Southern Depression has been completely infilled by flows originating from cones on the northern part of the crater floor, and the former prominent Saddle that separated the active part of the crater and the Southern Depression (see Figs. 9 and 10) now only exists as small ridges ascending from the crater floor (points labelled *S*). Since November 1988, the level of the crater floor has risen to the extent that the small cone on the inner north-western wall of the crater (point *X* on Fig. 10) is now almost buried (point *X*). The crater floor is around 30–40 m below the lowest point on the north-western rim of the crater. (Photo N. Patridge)

which by July 1992 became fully linked with the active part of the crater (Fig. 11). By 23 February 1993, the crater floor had risen to a point around 30 m below the eastern crater rim (Nyamweru 1993a).

Small lava effusions continued till June when, following an earthquake, ash eruptions were observed between 14 and 26th June, and Maasai herdsmen moved their cattle from the vicinity of the volcano. During an attempted ascent on 15 June, a local guide reported lava flowing down a gully high on the western slope. During an overflight in this period, large effusions of lava and lava fountaining were observed in the crater (M. Borner, pers. comm. to C. Nyamweru). A party from St. Lawrence University, Canton, New York, was on the volcano from 27–29 June, and observed the results of ash eruption and lava extrusion in the southern part of the crater, an area formerly free from activity. Their observations, together with analysis of photographs taken by C. Nyamweru on 3rd July, suggest the following sequence of events during the second half of June: (1) building of an asymmetric (perhaps partly collapsed) ash cone at the

Fig. 12. View south across crater, 3 July 1993. The recent activity is confined to the southern part of the crater, which is now the highest part of the crater floor. An asymmetric (?partly collapsed) ash cone is piled against the southern cliffs, and new whitened ash from this cone covers the summit area and the outer south-eastern slopes. A subsequent large, blockly lava flow from this cone covers most of the south-western part of the crater floor. Three smaller new ash cones penetrate the asymmetric cone. A blocky lava flow, piled up against the most easterly of the two prominent, older lava pinnacles (*foreground*), originates in the collapsed, crescent-shaped lava blister at the base of the east wall. Note that the crater floor is higher than in Figs. 10 and 11. (Photo C. Nyamweru)

base of the southern cliffs, with ash outfall on the outer slopes but most to the south and south-east (Fig. 12), (2) extrusion of a large, aa lava flow from this cone, covering most of the south-western part of the crater floor, (3) formation of three smaller ash cones penetrating the large ash cone, and covering its surface with new ash, (4) extrusion of lava from another vent close to the foot of the eastern wall, and (5) continued lava fountaining (observed on 29 June) and minor ash eruption from the more northerly of the three minor ash cones (Nyamweru 1993b). The surface of the eastern flow is slabby and, in this respect and in the large volume of both the flows, they are unlike any previously observed flows and require further investigation.

A.L. Tesha of the Tanzania Ministry of Mines and Resources reports ash eruptions and earth tremors during the latter half of July 1993 (pers. comm. to J.B. Dawson).

Note added in proof:
Despite their high viscosity, effusion rate and volume, the June 1993 lavas are dominantly of carbonatite, though of an exceptionally high crystallinity. They do, however, contain blebs of immiscible silicate material that, them-

selves, contain spheroids of carbonatite, thus providing firm evidence for the co-existence of silicate and carbonate liquids at the volcano (Dawson et al. 1994).

Acknowledgements. We wish to acknowledge the friends and colleagues, many mentioned in the text, who generously sent news and photographs of the activity at Oldoinyo Lengai. In particular, the Peterson brothers of Arusha have been the source of many recent reports. Their help was invaluable in the organization of the Kraff-Keller and Dawson expeditions in 1988, and in subsequent visits of others who have provided information about the volcano.

References

Barns TA (1921) The Highlands of the Great Craters, Tanganyika Territory. Geogr J 58: 401–416

Barns TA (1923) Across the Giant Craterland to the Congo. Ernest Benn, London, 271 pp

Dawson JB (1962a) The geology of Oldoinyo Lengai. Bull Volcanol 24:348–387

Dawson JB (1962b) Sodium carbonate lavas from Oldoinyo Lengai, Tanganyika. Nature 195: 1075–1076

Dawson JB (1992) Neogene tectonics and volcanicity in the north Tanzania sector of the Gregory Rift Valley: contrasts with the Kenya sector. Tectonophysics 204:81–92

Dawson JB (1993) A supposed sövite from Oldoinyo Lengai, Tanzania: result of extreme alteration of alkali carbonate lava. Miner Mag 57:93–101

Dawson JB, Bowden P, Clark GC (1968) Activity of the carbonatite volcano Oldoinyo Lengai, 1966. Geol Rundsch 57:865–879

Dawson JB, Garson MS, Roberts B (1987) Altered former alkalic carbonatite lava from Oldoinyo Lengai, Tanzania: inferences for calcite carbonatite lavas. Geology 15:765–768

Dawson JB, Smith JV, Steele IM (1989) Combeite ($Na_{2.33}Ca_{1.74}others_{0.12}Si_3O_9$) from Oldoinyo Lengai, Tanzania. J Geol 97:365–372

Dawson JB, Pinkerton H, Norton GE, Pyle D (1990) Physicochemical properties of alkali carbonatite lavas: data from the 1988 eruption of Oldoinyo Lengai, Tanzania. Geology 18:260–263

Dawson JB, Smith JV, Steele IM (1992) 1966 ash eruption of the carbonatite volcano Oldoinyo Lengai: mineralogy of lapilli and mixing of silicate and carbonate magmas. Miner Mag 56: 1–16

Dawson JB, Pinkerton H, Pyle, DM, Nyamweru C (1994) June 1993 eruption of Oldoinyo Lengai, Tanzania: exceptionally viscous and large carbonatite flows and evidence for co-existing silicate and carbonate magmas. Geology 22:799–802

Du Bois CGB, Furst J, Guest NJ, Jennings DJ (1963) Fresh natrocarbonatite lava from Oldoinyo L'Engai. Nature 197:445–446

Farler JP (1882) Native routes in East Africa from Pangani to the Masai country. Proc R Geogr Soc New Ser 4:730–753

Fischer GA (1885) Bericht über die im Auftrage der Geographischen Gesellschaft in Hamburg unternommene Reise in das Masai-Land 1882–1883. II: Begleitworte zur Original-Routenkarte. Mitt Geogr Ges Hamb 1885:189–237

Goriatschev AB (1968) Eruption of Oldoinyo Lengai. Priroda 1968, 7:88–92 (in Russian)

Guest NJ (1956) The volcanic activity of Oldoinyo L'Engai, 1954. Rec Geol Surv Tanganyika 4:56–59

Hay RL (1976) Geology of the Olduvai Gorge. University of California Press, Berkeley, 203 pp

Hay RL (1989) Holocene carbonatite-nephelinite tephra deposits of Oldoinyo Lengai, Tanzania. J Volcanol Geotherm Res 37:77–91

Hobley CW (1918) A volcanic eruption in East Africa. J East Afr Uganda Nat Hist Soc 13: 339–343

Keller J, Krafft M (1990) Effusive natrocarbonatite activity of Oldoinyo Lengai, June 1988. Bull Volcanol 52:629–645

Krafft M, Keller J (1989) Temperature measurements in carbonatite lava-lakes and flows: Oldoinyo Lengai, Tanzania. Science 245:168–190

MacIntyre RM, Dawson JB, Mitchell JG (1974) Age of fault movements in the Tanzania sector of the East African Rift system. Nature 247:354–356

Mittelholzer W (1930) Kilimandscharo-Flug. Orell Füssli, Zürich, 114 pp

Nyamweru C (1988) Activity of Ol Doinyo Lengai volcano, Tanzania, 1983–1987. J Afr Earth Sci 7:603–610

Nyamweru C (1990) Observations on changes in the active crater of Ol Doinyo Lengai from 1960 to 1988. J Afr Earth Sci 11:385–390

Nyamweru C (1993a) Ol Doinyo Lengai (Tanzania): increased carbonatitic lava production; ash eruptions. Smithonian Institution (30 April 1993). Glob Volc Network Bull 18(4):15–16

Nyamweru C (1993b) Ol Doinyo Lengai (Tanzania): description of crater in early July. Smithonian Institution (31 Aug 1993). Glob Volc Network Bull 18(8):4–5

Reck H (1914) Oldoinyo L'Engai, ein tätiger Vulkan im Gebiete der Deutsch-Ostafrikanischen Bruchstufe. Branca-Festschrift. Bornträger, Leipzig, pp 373–409

Reck H (1924/1925) L'Engai Bilder. Z Vulk 8:172–174

Reck H, Schulze G (1921) Ein Beitrag zur Kenntnis des Baues und der jüngsten Veränderung des L'Engai Vulkanes im nördlichen Deutsch-Ostafrika. Z Vulk 6:47–71

Richard J (1942) Volcanological observations in East Africa. I Oldoinyo Lengai. The 1940–41 eruption. J East Afr Uganda Nat Hist Soc 16:89–108

Uhlig C (1905) Bericht über die Expedition der Otto Winter-Stiftung nach den Umgebungen des Meru. Z Ges Erdkunde 1905:120–123

Uhlig C (1907) Der sogenannte Grosse Ostafrikanische Graben zwischen Magad (Natron See) und Llawa ya Mueri (Manyara See). Geogr Z 15:478–505

Uhlig C, Jaeger F (1942) Die Ostafrikanische Bruchstufe und die angrenzenden Gebiete zwischen den Seen Magad und Lawa ja Mweri sowie dem Westfuss des Meru. Deutsches Inst für Länderkunde, Wiss Veröff New Ser 10, 284 pp

Wakefield T (1870) Routes of native caravans from the coast to the interior of Eastern Africa, chiefly from information given by Sadi Bin Ahedi, a native of a district near Gazi, in Udigo, a little north of Zanzibar. J R Geogr Soc 40:303–338

Wichmann N (1894) Geographischer Monatsbericht (Afrika). Petermans Mitt 40:271–272

Williams RW, Gill JB, Bruland KW (1986) Ra-Th disequilibrium systems: Timescale of carbonatite magma formation at Oldoinyo Lengai volcano, Tanzania. Geochim Cosmochim Acta 50:1249–1259

Field Observations and Measurements of the Physical Properties of Oldoinyo Lengai Alkali Carbonatite Lavas, November 1988

H. Pinkerton[1], G.E. Norton[2], J.B. Dawson[3], and D.M. Pyle[4]

Abstract

Rheological and thermal properties of natrocarbonatite lavas from Oldoinyo Lengai have been measured in the field and in the laboratory. The viscosity, thermal diffusivity, specific heat capacity and latent heat of fusion are considerably lower than in basaltic lavas. A comparative study has revealed that lava flows erupted on Oldoinyo Lengai are smaller, though morphologically similar to basaltic lavas erupted on Mount Etna and Hawaii. Factors governing flow lengths, morphology and the development of compound flow fields on Oldoinyo Lengai are identical to those on basaltic volcanoes.

1 Introduction

Before the June 1988 expedition to Oldoinyo Lengai (Krafft and Keller 1989; Nyamweru 1989a; Keller and Krafft 1990) there were no direct measurements of the physical properties of natrocarbonatite lavas. Previous estimates of their physical properties were based on observations of the morphology of carbonatite lavas extruded in 1960 (Dawson 1962) and comparisons with synthetic alkali carbonate melts (Treiman and Schedl 1983; Treiman 1989).

One of our objectives during the visit to Oldoinyo Lengai in November 1988 (Dawson et al. 1990) was to perform a series of measurements that would complement and extend those made in June, 1988 (Dawson 1962; Krafft and Keller 1989; Keller and Krafft 1990). We also wished to establish whether the eruption temperatures and related thermal properties changed between June and November 1988. Field and laboratory measurements of the rheological properties of natrocarbonatites were made because previous observers (Krafft and Keller 1989; Keller and Krafft 1990) had commented on the low viscosities of the Oldoinyo Lengai lavas.

[1] Environmental Science Division, Institute of Environmental and Biological Sciences, Lancaster University, Lancaster LA1 4YQ, UK
[2] British Geological Survey, Keyworth, Nottingham NG12 5GG, UK
[3] Department of Geology and Geophysics, University of Edinburgh, West Mains Road, Edinburgh EH9 3JW, UK
[4] Department of Earth Sciences, Cambridge University, Downing Street, Cambridge CB2 3EQ, UK

We also compared the eruptive activity on Oldoinyo Lengai with effusive activity on basaltic volcanoes. In view of the small size and consequent slow advance rates of many lava flows, Oldoinyo Lengai is a perfect laboratory for studying the development of lava flows. Detailed observation of these flows support current flow-field models which have been developed for basaltic volcanoes. However, we also draw attention to other processes, such as thermal erosion, that have not been studied on basaltic volcanoes.

2 Changes in the Crater Between June and November, 1988

Several changes had taken place between June and November 1988 (see Nyamweru 1989a,b and Dawson et al., Chap. 4, this Vol., for a detailed account). The three lava lakes which were active during the June expedition (Centres 4, 7 and 5 in Fig. 1) had been replaced by spatter cones, and a new vent, that formed 10 h before the June expedition left the summit (Centre 8), had stopped erupting. Centre 9, which had formed near the eastern wall of the crater in June, and Centre 10 both became inactive in October. When we first arrived at the summit at 08.00 h on November 22, the activity was Strombolian

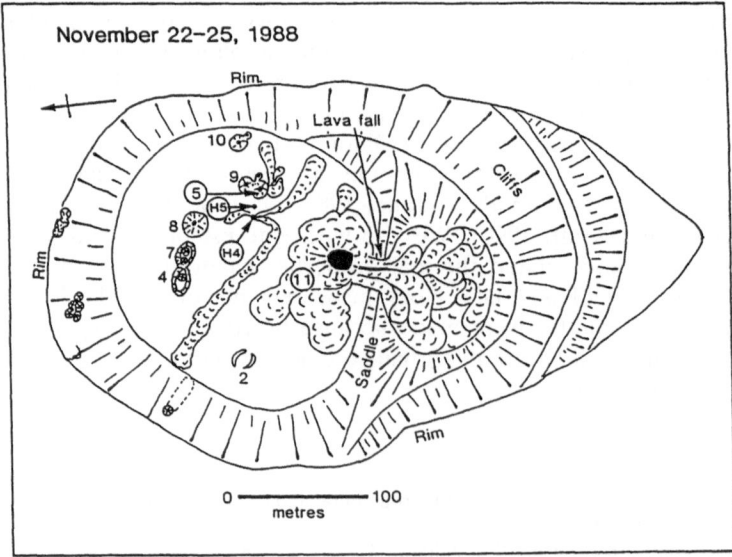

Fig. 1. Sketch map of the active crater of Oldoinyo Lengai to show the flows extruded in the period 22–25 November, 1988. Centre 11 is a breached spatter cone which was the source of lava which flowed down the 2-m-deep spillway channel during the first 2 days spent on the volcano. *Numbers* refer to centres reported in SEAN Bulletin 13 (1988) by C. Nyamweru. Centres active at the time of the November expedition are *circled*. Symbol *H* denotes a hornito. Crater flows erupted prior to 22 November 1988 are not shown

at the main vent, Centre 11, while degassing was taking place at a number of other vents.

2.1 Centre 11

Centre 11 (Fig. 1) was the main centre of activity within the North Crater, and spatter was ejected to heights of 10 m when large gas bubbles rose to the surface of the 8 m diameter lava pond. The mean repose time between eruptive events was 1.5 s, though this varied from one to several seconds. During our 3-day observation period, a crust formed on the outer part of the pool, decreasing its active diameter by 50%. This resulted in the formation of a pair of nested cones which were similar to those we observed at other inactive centres. During November 22 and 23, the main cone was breached on the southern side, and lava flowed out of the spillway down a 1-m-deep channel across the saddle (Fig. 1) and cascaded 5 m down a lava fall into the Southern Depression. Lava extrusion came in surges, sometimes overspilling the channel, with maximum effusion rates of $0.3 \, m^3 \, s^{-1}$. The advancing flows glowed incandescently at night, as did lava in Centre 11. Incandescence was also noted during the 1940–41, 1954 and June 1988 eruptions (Krafft and Keller 1989). By contrast, the 1960 lavas were not incandescent (Dawson 1962), suggesting that they may have been erupted at a lower temperature. Our observations indicate that activity at this vent was similar to Centre 7 in June (Keller and Krafft 1990).

2.2 Other Centres

During November 22, only one other vent apart from Centre 11 showed any sign of activity. The "Rhino vent" (Centre 4) was degassing in pulses (approximately 1 event/s) and was accompanied by a characteristic roar. The maximum gas temperature measured at this vent was 482 °C. During the night of November 22/23, the Rhino vent ejected clinker, up to 70 mm in diameter, a distance of 10 m from the vent, and a flow, approximately 150 m long, was erupted from a small hornito (Hornito 4). By 05.00 h on November 23, the level of lava in Centre 11 had dropped by 1 m and by 08.00 h on November 23, the overflow channel from Centre 11 had blocked; no lava flowed down the overflow channel during the remainder of our observation period.

On November 24th we witnessed the formation of a new hornito at the base of Centre 5. Small, gas-rich flows began to effuse near the base of Centre 5, gradually building up a small hornito (Hornito 5), with an internal diameter of 1 m (Fig. 2). Effusion rates, which were determined by measuring the dimensions and advance rates of small pahoehoe flows, varied from 0 to $8 \times 10^{-3} \, m^3 \, s^{-1}$ during the first 30 min of the life of Hornito 5. Five minutes later, there was a noise of blocks falling inside the cone at Centre 5 and by 05.45 a 1-m hole had formed. Lava was erupted through this opening, and 4 h later a number of small flows were being erupted from the southwestern base of Centre 5. These were typical gas-poor lavas, with a mean effusion rate, during the first 12 h, of

Fig. 2. Hornito 5 filled with frothy lava. The rheological properties of the lava are being measured using the rotating vane viscometer described in the text

$2 \times 10^{-3}\,\mathrm{m^3\,s^{-1}}$. Low lava levels in Hornito 5 corresponded to periods of high discharge rates from the base of Centre 5, suggesting that the vents were connected and that Hornito 5 was the degassing centre for Centre 5. A compound pahoehoe flow field was constructed around the base of Centre 5 during the remaining period of observation on the volcano.

3 Lava Flow Morphology

The most striking features of the flows on Oldoinyo Lengai were their small dimensions (flows were typically a few cm thick on eruption, and the maximum flow lengths were less than 200 m) and their low viscosity. In other respects the mode of development of the flow fields was very similar to typical basaltic compound flow fields on Mount Etna (Walker 1971; Pinkerton and Sparks 1976).

Two main types of lava erupted during the period of observation. One was gas-rich, with an estimated bubble content of 50%. The upper surface of these flows was glassy in appearance, and a considerable amount of gas loss was observed during flow. The other type contained less than 10% of bubbles and it was considerably more fluid than the gas-rich lavas. The main morphological differences between these two types of lava are that the gas-rich lava formed mainly thick, slow-moving aa and slab lava flows whereas the gas-poor lava formed thin slab and pahoehoe flows.

Aa lavas formed on high-effusion rate $(0.2-0.3\,\mathrm{m^3\,s^{-1}})$, gas-rich flows on Oldoinyo Lengai, whereas pahoehoe textures formed on gas-rich flows when effusion rates were less than $0.04\,\mathrm{m^3\,s^{-1}}$. Slab lava formed at the fronts of higher-effusion-rate pahoehoe flows, and the slabs became fragmented if the flow was subjected to variations in discharge rates or if the flow widened on reaching a lower gradient. Toothpaste lava was observed at the front of two stationary flows. This represents the final effusion of cool, plastic lava from the interior of cooling flows (Rowland and Walker 1987). Another type of lava that was erupted from cracks in the flow fronts flow fronts was, by contrast, a very fluid aphyric lava, as described by Keller and Krafft (1990). In common with basaltic flows (Walker 1973; Pinkerton 1987), the maximum cooling-limited lengths of individual carbonatite flows were related to effusion rate. The maximum length attained by a flow during the observation period was 150 m.

The formation of the compound flow field on Oldoinyo Lengai was complicated by frequent variations in discharge rates at the source vents. When supply rates to a flow decreased, a crust formed rapidly on top of the lava in the central channel. The crustal thickening rates decreased from 2 mm/min during the first 3 min on thin pahoehoe flows to 0.3–0.7 mm/min during the next 20 min. This cooled crust propagated upstream, eventually causing the vent region itself to become roofed over. If effusion rates now increased, the central channel and the area around the vent bulged, and the cooled slabs within the channels were uplifted. Within a few minutes, the upper surface of a smooth channelled flow was transformed into a jumble of slabs up to 10 mm in thickness and 1 m in length, inclined at up to 45 degrees from the horizontal. Lava would then break out, either within the flow or upstream of the old vent. This cycle was repeated at some vents several times.

Channels on Oldoinyo Lengai varied widely in size and in their mechanisms of formation. On eruption, flows of new, gas-poor, incandescent lava spread out rapidly as thin, unconfined flows, and levees formed only when the flow margins had visibly cooled. New lava was then constrained to flow between the cooled levees. By contrast, gas-rich lavas did not spread out as unconfined flows on eruption. Instead, they had well-developed levees and channels as soon as they were erupted from the vent, in spite of their greater thicknesses and hence the reduced effects of cooling. By analogy with basaltic lava flows, these levees appear to form as a consequence of the non-Newtonian properties of the gas-rich lavas, and the levee and channel dimensions are dictated by a Bingham flow model (Hulme 1974; Sparks et al. 1976).

Our observations of channel development suggest that uncooled, gas-poor, carbonatite lavas have viscosities which are lower than basaltic lavas, and they do not possess a significant yield strength. In contrast, gas-rich lava flows either possess a yield strength on eruption, or they are pseudoplastic. The validity of these conclusions is investigated using rheological measurements which we present later.

During measurements of velocities and effusion rates of the active carbonatite flows, it was observed that, once flow was established in these channels for a few minutes, the depths of the channels increased. Systematic measurements of channel dimensions (using a tape measure inserted into the active flows) and flow velocities revealed that the substrate was being eroded, even when it was fresh, unfissured pahoehoe. The velocities of lava in the central upper part of channels were calculated by noting the mean time required for mm-sized fragments of ash place on the surface of the flow to travel a distance of 1 m. A minimum of six velocity measurements were made at each locality. Erosion rates and amounts decreased systematically downstream even though there were no temporal or spatial changes in effusion rate. Measurements of the increase in depth of a 0.13-m-deep and 0.21-m-wide channel at a distance of less than 1 m from its source indicated that the erosion rate remained constant, over the 9-min observation period, at 2 mm/min. The velocity of lava in this channel was 0.25 m/s. Careful measurements and observations of these small channels confirmed that the primary erosion mechanism was by melting, although the presence of 0.17 m of partially solid mush at the base of the 1.1 m channel below the Centre 11 cascade suggests that, once some of the substrate beneath a channel had melted, the residual crystalline material was detached by the high shear stresses at the base of the channel. In plan view (Fig. 3), many of the resulting channels are remarkably similar to lunar sinuous rilles.

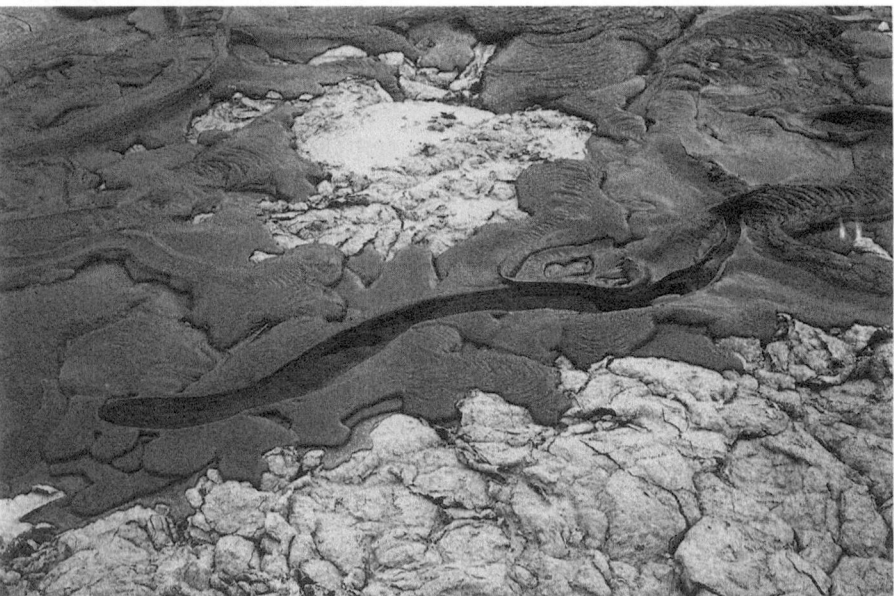

Fig. 3. This small sinuous channel is typical of many which formed on Oldoinyo Lengai during late November, 1988. The depth of this channel increased at a rate of 2 mm/min due to thermal erosion under laminar flow conditions. The channel had a maximum width of 150 mm

It has been suggested that lunar sinuous rilles may have been formed by substantial melting of the substrate (Hulme 1982; Wilson et al. 1987). Huppert et al. (1984) and Barnes et al. (1988) have also postulated that thermal erosion played an important role in the ascent and eruption of komatiite magmas. For both lunar basalts and komatiites, the flow regime is considered turbulent. However, calculated Reynolds numbers for lava within the channels on Oldoinyo Lengai were typically two orders of magnitude lower than the value at which turbulence would begin. Our observations on Oldoinyo Lengai are important because this is the first time that thermal erosion has been measured under laminar flow conditions.

Models of the emplacement and thermal erosion of carbonatite lava flows require accurate measurements of their physical properties. Since estimates of thermal and rheological properties by Treiman and Schedl (1983) were based on comparisons with the properties of synthetic alkali carbonate melts at temperatures over 500 °C higher than those measured on Oldoinyo Lengai, and since, as we show later, the viscosity of carbonatites change by up to three orders of magnitude over this temperature range, their estimates cannot be used to model volcanic processes on this volcano. Measurements of the thermal and physical properties of lavas were therefore measured in the field, and additional measurements were performed on samples in our laboratory.

4 Field and Laboratory Measurements

4.1 Temperature and Cooling Rates of Flows

Lava temperatures were measured with a self-compensating Comark Cr-CrAl thermocouple that was calibrated against the melting points of Sn, Zn and Al (Norton 1991). Temperatures were consistently in the range 576–593 °C. There was no apparent difference in the eruptive temperatures of the gas-free and gassy lavas. The highest temperatures (593 °C) were measured at flow fronts and in the top 70 mm of the lava pool in Hornito 5 (Fig. 2). These temperatures are higher than the maximum temperature of 544 °C recorded in June 1988 (Krafft and Keller 1989). The absence of incandescence during the 1960 activity and the lower temperatures measured in June 1988 (Krafft and Keller 1989) suggest variation in eruptive temperatures by tens of degrees Centigrade.

The thermocouple was also used to determine cooling rates at the centre of a small gas-poor pahoehoe flow that had been erupted onto cool pahoehoe flows (Fig. 4). Using these measurements, the thermal diffusivity of the carbonatites can be calculated using a solution of Fourier's equation once the specific heat capacity and latent heat of crystallization are known. Previous estimates of these properties for carbonatites are based on comparison with a pure synthetic sodium carbonate melt at 827 °C and 1 bar pressure.

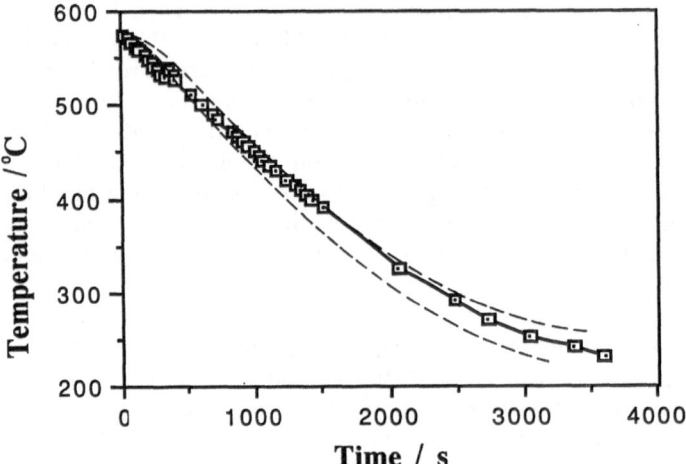

Fig. 4. Measured and calculated cooling rates at the centre of a 31-mm-thick natrocarbonatite lobe of pahoehoe lava erupted from Centre 5 on Oldoinyo Lengai. The *dashed curves* are calculated using the equations of Carslaw and Jaeger (1959). The higher calculated trend is for a thermal diffusivity of $4 \times 10^{-8} \, m^2 s^{-1}$ and the lower trend is for a diffusivity of $5 \times 10^{-8} \, m^2 s^{-1}$. The jump in the measured cooling curve is due to a new influx of lava close to the point of measurement

4.2 Measurements of Specific Heat Capacity and Latent Heat of Crystallization

The lavas we collected are chemically very similar to one another (Dawson et al. 1990; Keller and Krafft 1990) and the small variations can be explained largely by differing phenocryst contents (Dawson et al., this Vol.). The two major phenocryst phases distinguished petrographically are nyerereite and gregoryite. These phases are also present as microphenocrysts in the groundmass together with sylvite and minor phases (see Dawson et al., Chap. 4, this Vol.).

A differential scanning calorimeter (DSC) was used to measure specific heat capacity by comparing the heat flow into and out of a sample with that required to keep the temperature of the sample at the same value as a standard of known heat capacity. Using the mass and specific heat capacity of the standard and the mass of the sample, the specific heat capacity of the sample can be calculated (O'Neill 1966). The specific heat capacities of four carbonatite samples have been measured at temperatures between 47 and 567 °C. Measurements cannot be made close to phase transitions in the sample since these produce anomalous peaks on the DSC plot. The heats of reactions can be calculated from the DSC plot by dividing the area under the reaction peak by the sample mass. The measured specific heat capacities and latent heats of fusion of three of the carbonatite samples are presented in Table 1, where it can be seen that there is very good agreement between our value of latent heat of fusion for BD4155, the aphyric natrocarbonatite, and that estimated by Treiman and Schedl (1983). If

Table 1. Physical properties of powdered samples of three natrocarbonatite lavas collected during November 1988 compared with the values estimated by Treiman and Schedl (1983).[a] BD4155 is a non-vesicular, glassy lava that was sampled from an overspill of a channel flow to the south of Centre 11; BD4160 is a non-vesicular, phenocryst-poor lava from the centre of the flow from Centre 5 on which the thermal diffusivity measurements were made; and BD4169 is a highly vesicular, phenocryst-rich specimen from Hornito 5. All samples were collected and sealed in polythene bags as soon as they had cooled. Further descriptions of the samples are given in Dawson et al. (1990 and Chap. 4, this Vol.) and Norton (1991). The lavas have similar major element chemistries and the small variations are attributed to differing phenocryst contents (Dawson et al. Chap. 4, this Vol.)

Sample no.	Thermal diffusivity	Specific heat capacity[b]	Density at 20 °C	Temperature on eruption	Latent heat of fusion[c]	Apparent viscosity on eruption
	$/m^2 s^{-1}$	$/J kg^{-1} K^{-1}$	$/kg m^{-3}$	$/°C$	$/kJ kg^{-1}$	$/Pa s$
BD4155	nd	1095–1223	2170	583	119–122	nd
BD4160	4×10^{-8}	1043–1307	nd	586	82–102	0.6–5.3
BD4169	nd	1064–1485	1150	592	58–59	14–73
Treiman and Schedl		2000	2200	nd	125	5×10^{-3}

[a] Treiman and Schedl's (1983) estimates of latent heat of fusion, specific heat capacity and density were made at 1100 K.
[b] The first value in this column is for a temperature of 77 °C; the second is for 397 °C.
[c] The two values in this column are for duplicated measurements on different powders from the same sample.

our values of specific heat capacity are extrapolated to 1100 K, our results are similar to the value of $2000 J kg^{-1} K^{-1}$ predicted by Treiman and Schedl (1983).

4.3 Field Measurement of Thermal Diffusivity

In one small pahoehoe lobe from Centre 5, the rate of cooling of the flow was measured by inserting the thermocouple into the centre of the flow after it had stopped moving. The tip of the thermocouple was inserted 114 mm into the flow front. Since the flow was only 90 mm wide and 31 mm deep, conduction of heat along the length of the thermocouple was small relative to conduction of heat through the lava perpendicular to the thermocouple. The rate of cooling of this flow is shown in Fig. 4. Theoretical temperature-time curves can be generated for lavas with different thermal diffusivities using the solution of Fourier's heat conduction equation in Carslaw and Jaeger (1959, p. 58) Laboratory measurements of specific heat capacity and latent heat of crystallisation are used in these calculations. The resulting curves can then be compared with the measured curve, and the appropriate thermal diffusivity can be determined (Fig. 4). A two-dimensional treatment of heat flow can be justified because the flow width was three times the flow thickness. It can be seen that, between the solidus (480 °C) and 380 °C, the thermal diffusivity is $4 \times 10^{-8} m^2 s^{-1}$. As the temperature drops to 250 °C, the diffusivity increases to $5 \times 10^{-8} m^2 s^{-1}$. These values are over an order of magnitude smaller than the thermal diffusivity of basaltic

Table 2. Channel velocities, depths and gradients required to estimate the apparent viscosities of lava within small channels on Oldoinyo Lengai

Mean velocity /ms^{-1}	Gradient /degrees	Flow depth /m	Apparent viscosity /Pas
1.71	2.50	0.93	43
1.43	2.50	0.93	52
0.85	2.50	0.85	73
1.08	2.50	0.85	57
0.37	8.20	0.12	27
0.36	4.00	0.12	14
0.73	8.20	0.06	3.5
0.41	4.00	0.06	2.6
0.22	4.00	0.02	0.6
0.38	5.00	0.04	1.8
0.81	5.00	0.10	5.3

lavas (Pinkerton and Sparks 1976); thus these carbonatite lavas will cool considerably more slowly than basaltic lavas of comparable dimensions. Vesicles in the frothy flows further reduce thermal diffusivity.

4.4 Rheology

Two methods were used to determine the rheological properties of the carbonatite lava flows in the field. The first involved a portable rotating shear vane (Fig. 2) based on earlier versions used on Hawaii (Shaw et al. 1968) and Mount Etna (Pinkerton and Sparks 1976; Guest et al. 1987). The rotating shear vane is driven by a 24-V DC motor with a speed controller coupled through a torque meter to a shear vane. The system measures the torque required to rotate the vane at different rates, simultaneous rotation speeds being monitored using an optical tachometer. From the simultaneous measurements of torque and rotation rate, the rheological properties of the carbonatite lavas over a range of strain rates can be calculated (Spera et al. 1988). The second method involved measuring the slope, velocity, channel depth and densities of lavas flowing down active channels. Using these measurement, apparent viscosities can be calculated using Jeffreys' equation (Williams and McBirney 1979). The results are shown in Table 2. The non-Newtonian characteristics of the gas-rich flows can be seen in Fig. 5, where the mean strain rates are approximated by the velocity/depth ratio.

The rheological measurements show that the gas-free pahoehoe natro-carbonatite lavas (Fig. 3) had apparent viscosities in the range 1–5 Pas and they behaved as inelastic Newtonian fluids. By contrast, the frothy lava in the Hornito 5 lava pool (Fig. 2) had apparent viscosities of 70–120 Pas at low strain rates, but 0.3–3 Pas at high strain rates. The vesicular lava is therefore a

Fig. 5. Apparent viscosities of gas-rich lava flows erupted on Oldoinyo Lengai. Mean strain rates assume a linear decrease in velocity with depth. From this graph, it is clear that the gas-rich lavas from Oldoinyo Lengai are non-newtonian

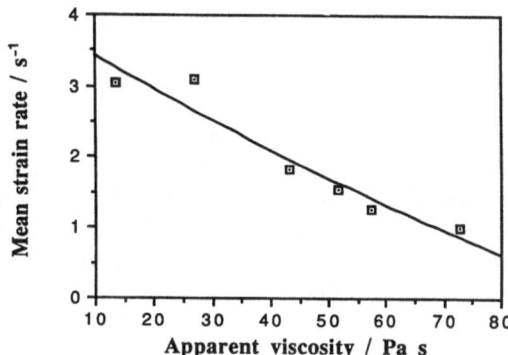

pseudoplastic froth. Pseudoplasticity in froths is caused by surface tension which prevents bubble deformation at low shear stresses, thereby increasing the apparent viscosity. However, at high strain rates the bubbles deform, and the apparent viscosity decreases. The gas-free carbonatites are therefore more than an order of magnitude less viscous than the most mobile basaltic lavas measured to date (Moore 1987).

In view of the current interest in the migration of carbonatite melts (Wallace and Green 1988; Bailey 1989), further measurements of the rheological properties of carbonatite melts have been made in our laboratory using a Haake Rotovisco rotating viscometer. The rheological properties of the natrocarbonatite lavas at different temperatures over a range of shear strain rates have been measured, and we have studied the effects of crystals and bubbles on carbonatite rheology. The high temperature, concentric cylinder rheological system used in this study is described by Norton (1991). All measurements were made in an oxygen-free nitrogen environment at a pressure of 1 bar. Preliminary interpretation of the results confirm that field values are close to those measured in the laboratory. These values also show that, if the viscosities measured in the laboratory are extrapolated to the higher temperatures used by Treiman and Schedl (1983) and Treiman (1989), they support their estimates. Their viscosity of 5×10^{-3} Pa s for natrocarbonatite is based on a comparison with a sodium carbonate melt at 827 °C and 1 bar pressure.

In conclusion, our measurements confirm that the natrocarbonatite lavas from Oldoinyo Lengai have uniquely low magmatic viscosities, and that their rheological properties on eruption depend largely on the vesicularity.

5 Conclusions

We have made the first field measurements of the rheological properties and thermal diffusivity of natrocarbonatite lavas. The temperatures measured during November, 1988 are 50 °C higher than those measured in June 1988 (Krafft and Keller 1989) and higher than the inferred temperatures during the 1960 activity.

This indicates that the eruptive temperatures of these carbonatite lavas vary by tens of degrees Centigrade. Our discharge measurements indicate that effusion rates were not constant during our observation period. In addition, the gas contents of the lava varied, both spatially and, for the gas-rich vents, with time.

The measured effusion temperatures of the lavas, apparent viscosities, thermal diffusivities, specific heat capacities and flow dimensions of natrocarbonatite lavas from Oldoinyo Lengai are significantly lower than typical values for basaltic lavas. Our studies confirm that, while estimates of the viscosity and specific heat capacity of synthetic alkali carbonatite melts (Treiman and Schedl 1983) appear to be at variance with our measurements on natrocarbonatites on Oldoinyo Lengai, this is because of the different temperatures which have been used. When our measurements of specific heat capacity and viscosity on aphyric natrocarbonatites are extrapolated to the higher temperatures used by Treiman and Schedl (1983), the results are in agreement. Similarly, the values for latent heat of fusion and density predicted by Treiman and Schedl (1983) are close to those we measured.

The lavas on Oldoinyo Lengai are smaller, though in other ways similar, morphologically, to basaltic lava flows. They are remarkably similar in size and in flow behaviour to natural and industrial flows of molten sulphur (Greeley et al. 1990). The main morphological difference between carbonatite and basaltic flow fields appears to be the ubiquitous development of thermal erosion channels in the proximal parts of carbonatite flows. Heat transfer equations which have been developed for laminar flow (Pinkerton et al. 1990) confirm that measured and calculated thermal erosion rates are similar and they predict that thermal erosion will also take place during many terrestrial basaltic eruptions. Difficulties in measuring lava depths in active channels may explain why this process is not generally recognized. Oldoinyo Lengai is therefore an ideal laboratory on which one can study the development of carbonatite flows and make observations and measurements which further our knowledge of both carbonatite and basaltic volcanoes.

Acknowledgements. The expedition to Oldoinyo Lengai was sponsored by the Royal Society and GEN was supported by the Natural Environment Research Council. We thank Celia Nyamweru, the Peterson brothers and Dorobo Safaris for their invaluable assistance in East Africa. Finally, we thank K. Bell, J.E. Guest, G. Mahood, B.O. Mysen, F.J. Spera and A.H. Treiman for their constructive criticisms on an earlier version of the manuscript.

References

Bailey DK (1989) Carbonate melts from the mantle in the volcanoes of South-East Zambia. Nature 338:415–418

Barnes SJ, Hill RET, Gole MJ (1988) The Perseverance ultramafic complex, Western Australia: the product of a komatiite river. J Petrol 29:305–332

Carslaw HS, Jaeger JC (1959) Conduction of heat in solids. Clarendon Press, Oxford, 510 pp

Dawson JB (1962) Sodium carbonate lavas from Oldoinyo Lengai, Tanganyika. Nature 195: 1075–1076

Dawson JB (1989) Sodium carbonatite extrusions from Oldoinyo Lengai, Tanzania: implications for carbonatite complex genesis. In: Bell K (ed) Carbonatites – genesis and evolution. Unwin Hyman, London, pp 255–277

Dawson JB, Pinkerton H, Norton GE, Pyle DM (1990) Physico-chemical properties of alkali carbonatite lavas: data form the 1988 eruption of Oldoinyo Lengai, Tanzania. Geology 18:260–263

Greeley R, Lee SW, Crown DA, Lancaster N (1990) Observations of industrial flows: implications for Io. Icarus 84:374–402

Guest JE, Kilburn CRJ, Pinkerton H, Duncan A (1987) The evolution of flow-fields: observations of the 1981 and 1983 eruptions of Mount Etna, Sicily. Bull Volcanol 49:527–540

Hulme G (1974) The interpretation of lava flow morphology. Geophys J R Astr Soc 39: 361–383

Hulme G (1982) A review of lava flow processes related to the formation of lunar sinuous rilles. Geophys Surv 5:245–279

Huppert HE, Sparks RSJ, Turner JS, Arndt NT (1984) Emplacement and cooling of komatiite lavas. Nature 309:19–22

Keller J, Krafft M (1990) Effusive natrocarbonatite activity of Oldoinyo Lengai, June 1988. Bull Volcanol 52:629–645

Krafft M, Keller J (1989) Temperature measurements in carbonatite lava lakes and flows from Oldoinyo Lengai, Tanzania. Science 245:168–170

Moore HJ (1987) Preliminary estimates of the rheological properties of the 1984 Mauna Loa lava. USGS Prof Pap 1350:1569–1588

Norton GE (1991) The physical properties of carbonatite and silicate magmas. PhD Thesis, Lancaster University, Lancaster (unpublished)

Nyamweru C (1988) Activity of Oldoinyo Lengai volcano, Tanzania, 1983–1987. J Afr Earth Sci 7:603–610

Nyamweru C (1989a) Report on activity in the northern crater of Oldoinyo Lengai, 24th June to 1st July 1988. J East Afr Nat Hist Soc Natl Mus 79(186):1–15

Nyamweru C (1989b) Report on activity in the northern crater of Oldoinyo Lengai, July 1988 to August 1989. J East Afr Nat Hist Soc Natl Mus 79(194):1–15

O'Neill MJ (1966) Measurement of specific heat functions by differential scanning calorimetry. Anal Chem 38:1331–1336

Pinkerton H (1987) Factors affecting the morphology of lava flows. Endeavour 11:73–79

Pinkerton H, Sparks RSJ (1976) The 1975 sub-terminal lavas, Mount Etna: a case history of the formation of a compound lava field. J Volcanol Geotherm Res 1:167–182

Pinkerton H, Wilson L (1992) The dynamics of channel-fed lava flows (abstract). Lunar Planet Sci 23:1083–1084

Pinkerton H, Wilson L, Norton GE (1990) Thermal erosion – observations on terrestrial flows and implications for planetary volcanism (abstract). Lunar Planet Sci 21:964–965

Rowland SK, Walker GPL (1987) Toothpaste lava: characteristics and origin of a lava structural type transitional between pahoehoe and aa. Bull Volcanol 49:631–641

Shaw HR, Wright TL, Peck DL, Okamura R (1968) The viscosity of basaltic magma: an analysis of field measurements in Makaopuhi lava lake, Hawaii. Am J Sci 226:225–264

Sparks RSJ, Pinkerton H, Hulme G (1976) Classification and formation of lava levees on Mount Etna, Sicily. Geology 4:269–271

Spera FJ, Borgia A, Strimple J, Feigenson M (1988) Rheology of melts and magmatic suspensions I. Design and calibration of concentric cylinder viscometer with application to rhyolitic magma. J Geophys Res 93, B9:10723–10294

Treiman AH (1989) Carbonatite magma: properties and processes. In: Bell K (ed) Carbonatites – Genesis and evolution. Unwin Hyman, London, pp 89–104

Treiman AH, Schedl A (1983) Properties of carbonatite magma and processes in carbonatite magma chambers. J Geol 91:437–447

Walker GPL (1971) Compound and simple lava flows. Bull Volcanol 35:579–590

Walker GPL (1973) Lengths of lava flows. Philos Trans R Soc Lond A274:107–118

Wallace ME, Green DH (1988) An experimental determination of primary carbonatite magma composition. Nature 335:343–346

Williams H, McBirney AR (1979) Volcanology. Freeman & Cooper, San Francisco, 397 pp

Wilson L, Pinkerton H, Macdonald R (1987) Physical processes in volcanic eruptions. Annu Rev Earth Planet Sci 15:73–95

The Dynamics of Degassing at Oldoinyo Lengai

D.M. Pyle[1], H. Pinkerton[2], G.E. Norton[3], and J.B. Dawson[4]

Abstract

This chapter summarizes field observations of degassing made at Oldoinyo Lengai in 1988 and their implications for the dynamics of the Lengai magma system. Observations of bubbles bursting in a lava lake, and sudden changes in magma level and vesicularity at active vents are consistent with the degassing model developed by Jaupart and Vergniolle (1988, 1989), and suggest that the shallow plumbing system comprises several small (<10-m radius) chambers containing transient foam layers which may, in turn, be connected at depth to a single, larger chamber.

1 Introduction

While phenomena such as the explosion of bubbles in lava lakes, the vigorous "boiling" of lava pools, and rapid temporal changes in lava levels have been observed at Lengai (Nyamweru 1988; Krafft and Keller 1989; Dawson et al. 1990; Keller and Krafft 1990), no attempt at understanding these phenomena or their implications has been made before now. Recent studies by Jaupart and Vergniolle (1988, 1989) bear on this problem, and allow interpretation of field observations of these phenomena. Here, we present the relevant field observations, and discuss their implications for the mechanism of degassing at Oldoinyo Lengai.

1.1 Outline of Activity, 1988

Oldoinyo Lengai has erupted carbonatite lava periodically since 1880 (Hobley 1918; Dawson 1962, 1989; Keller and Krafft 1990), with alternating periods of lava effusion, repose, and mildly explosive activity. During the current eruptive

---

[1] Department of Earth Sciences, University of Cambridge, Downing Street, Cambridge CB2 3EQ, UK
[2] Environmental Science Division, Institute of Environmental and Biological Sciences, University of Lancaster, Lancaster LA1 4YQ, UK
[3] British Geological Survey, Keyworth, Nottingham NG12 5GG, UK
[4] Grant Institute of Geology, University of Edinburgh, West Mains Road, Edinburgh EH9 3JW, UK

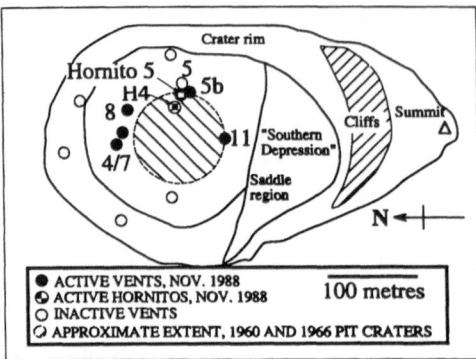

Fig. 1. Summary map of the crater in November 1988, with features numbered after Nyamweru (1988, 1989). The *hatched area* outlines the extent of the 1966 pit crater. (Dawson et al. 1968)

episode, lava has erupted semi-continuously since 1983 (Nyamweru 1988). This report refers to a visit made in November 1988. Results of other studies made during this visit have been reported elsewhere (Nyamweru 1989; Dawson et al. 1990; Pyle et al. 1991; Pinkerton et al., Chap. 2, this Vol.).

Figure 1 is a map of the crater floor at the time of observation. Centre 11 was a 10-m-diameter lava lake that intermittently disgorged lava through a 1-m-deep channel into the "Southern Depression". Centre 5 was a cluster of older vents, which had been active intermittently through the previous 2 years. Two new vents (Hornito 5, Centre 5b) were observed forming on 24/11/88, while elsewhere in the crater, magma was observed at depth at Centre 8, and fresh ejecta were observed at Centre 4/7 on November 24th.

2 Models of Degassing Magma Systems

Bubbles play a major role in the behaviour of near-surface magma bodies. Detailed modelling of aspects of the degassing behaviour of silicate magmas has been possible (e.g. Sparks 1978) because both physical (viscosity, density, temperature) and chemical (solubility, diffusivity) properties of melt and volatile phases are fairly well known. Similar models should apply to shallow degassing bodies of carbonatite magma, but remain unexplored. In this chapter we illustrate how observations of bubbles exploding in lava pools and rapid variations of magma level in surface vents can be explained as simple consequences of the storage and movement of bubbles and melt in the sub-surface. Existing models may be used to quantify the characteristic dimensions of the degassing sub-surface reservoirs at Oldoinyo Lengai.

2.1 The Behaviour of Magmatic Foam

Bubbles exsolved within a degassing magma body will rise buoyantly, accumulating at the roof as a foam layer. The subsequent behaviour of the foam layer controls the nature of the eruption through the conduit: the foam layer grows as more bubbles reach the chamber roof and thins as foam is transported out of the conduit. Above a critical foam thickness, bubbles in contact with the chamber roof will coalesce as surface tension forces are exceeded, and foam collapse will occur. The status of the foam layer depends primarily on the foam viscosity, the bubble size and the gas flux into the system (Jaupart and Vergniolle 1988, 1989; Vergniolle and Jaupart 1990). At high enough gas fluxes, the foam layer grows faster than it is removed from the system, leading to a cycle of foam layer growth, collapse and regeneration. When the foam is of low viscosity, collapse events transform the whole foam layer into single gas pockets which then escape from the conduit. This "cyclic" behaviour may correspond to that observed at Kilauea during fire-fountaining episodes (Vergniolle and Jaupart 1990). At higher foam viscosities, collapse is retarded, and beyond its critical thickness, the foam only partially collapses into small gas "slugs". Stromboli exhibits "slugging" behaviour (Jaupart and Vergniolle 1989).

The ratio of the characteristic timescales of gas pocket growth during coalescence to the residence time of that pocket in the foam layer, N_2 (Jaupart and Vergniolle 1989), determines whether the ensuing regime exhibits cyclic or slugging behaviour:

$$N_2 = \frac{\left\{\frac{2\mu_m a}{\sigma}\right\}}{(r^2) \cdot \left(\frac{3\pi^3 \mu_m \varepsilon^2}{Q^3 \rho_l \cdot g}\right)^{\frac{1}{4}}} \tag{1}$$

with parameters as defined in Table 1. The cyclic regime is characterized by small values of N_2 (<0.1; Jaupart and Vergniolle 1989). The repose period (t_c) between explosions is a function of the gas flux into the system (Q) and the radius of the flat top of the tank or magma chamber (r):

$$t_c = \pi r^2 \cdot \left\{\frac{2\sigma}{\rho_l g a Q}\right\}. \tag{2}$$

The critical thickness of foam (h_c) is the thickness above which the layer will collapse and coalesce, and is given approximately (Jaupart and Vergniolle 1989) by

$$h_c = \frac{2\sigma}{\varepsilon \rho_l g a}. \tag{3}$$

With appropriate characterization of the degassing rate and style, it should be possible to place constraints on the dimensions (r) of the storage region using this analysis.

Table 1. Summary of symbols and physical properties

Description	Symbol	Value and dimensions	Reference[a]
Gas volume fraction in the foam	ε	0.69	1
Exit angle of ejecta	ϕ	–	
Gas density	ρ_g	$kg\,m^{-3}$	
Magma density	ρ_l	$2170\,kg\,m^{-3}$	2
Bulk foam density	ρ_m	$680\,kg\,m^{-3}$	b
Magma surface tension	σ	$0.22\,N\,m^{-1}$	3[c]
Melt viscosity	μ_l	1 Pas	2
Bulk foam viscosity	μ_m	100 Pas	2[b]
Bubble radius	a	1 mm	
Conduit diameter	d	m	
Eötvos number	Eo		
Acceleration due to gravity	g	$9.81\,m\,s^{-1}$	
Foam layer thickness	h	m	1
Critical foam thickness	h_c	m	1
Morton number	Mo		
Gas flux into the magma system	Q	$m^3\,s^{-1}$	
Range of ejecta	R	m	
Reynolds number	Re		
Magma chamber radius	r	m	
Period between gas bursts	t_e	1–2 s	
Time of flight	t_f	s	
Ejecta velocity	U	$m\,s^{-1}$	
Maximum erupted velocity of ejecta	U_o	$m\,s^{-1}$	
Terminal bubble rise velocity	V_b	$m\,s^{-1}$	
Spherical cap base radius	y	m	

[a] 1, Jaupart and Vergniolle (1989); 2, Dawson et al. (1990); 3, Janz (1988).
[b] Assuming 69% vesicularity.
[c] Janz (1988) quotes the relationship between surface tension (σ) and temperature for 100% molten Na_2CO_3 at temperature T (K) as $\{268.5-0.0502\,T\}\,mNm^{-1}$, for T = 1143–1279 K. We allow for the effects of K_2CO_3, which lowers σ and assume that the data may be extrapolated to lower T; there are no data for the effects of Ca on σ.

3 Degassing at Oldoinyo Lengai: Observation and Interpretation

3.1 Centre 11

Two- to 3-diameter bubbles, measured by optical range-finder, exploded continuously within the lava lake at Centre 11 for the entire period of observation. The interval between explosions was consistently between 1.1 and 1.9 s (Fig. 2). Individual explosions scattered fresh ejecta 10–20 m into the air, with a horizontal range of <10–15 m and a time of flight of <2 s.

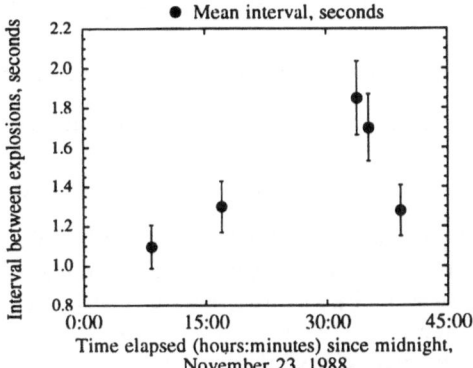

Fig. 2. Summary of observations of the mean interval between bubble explosions in the lava pond at Centre 11, November 23–24, 1988, based on the interval taken for between 20 and 145 explosions; 10% relative error bars are shown, based on the variability of seven repeat measurements in one time interval. Slugs of gas typically exploded every 1–2 s

3.1.1 Analysis of Bubble Explosions

The maximum velocity of ejecta from the lava lake explosions limits the excess pressure of the exploding bubbles. Ejection velocities of the scoriae may be estimated from simple time of flight equations, neglecting air resistance and drag. This approximation is valid, since the explosions at Oldoinyo Lengai were relatively small, and ejecta travelled only a very short distance at low velocities (cf. Wilson 1972). For a projectile the relationship between the expected horizontal range R and the time of flight is given (from standard formulae) by

$$\frac{R}{t_f^2} = \frac{g}{4 . \tan\phi},$$ (4)

so for ejection angles $\phi > 30°$ this parameter should be <4. At Lengai, $R/t^2 \sim 2.5–5$, confirming that it is appropriate to use these simple ballistic equations. Thus, for a projectile of velocity U_o the projectile range is given by

$$R = \frac{U_o^2 . \cos\phi . \sin\phi}{g},$$ (5)

where g is the acceleration due to gravity. For R < 10–15 m, and an optimal angle of 45°, this gives $U_o < 20\,\mathrm{m\,s^{-1}}$, confirming the low eruption velocities of the ejecta. Applying the equations of Wilson (1980) with suitable adjustments to allow for CO_2 bubbles in carbonatite magma (rather than water in a silicate melt), indicates that for 2–3-m diameter bubbles such ejection velocities could be attained for excess bubble pressures of $<2 \times 10^4\,\mathrm{Pa}$ (0.2 bar). This complements the conclusions of Blackburn et al. (1976) and Wilson (1980) that the driving force for Strombolian eruptions is small. A more thorough analysis of 1974 eruptions of Stromboli (Chouet et al. 1974) revealed similar ejecta velocities of 15–26 m s^{-1}.

3.1.2 Volumetric Degassing Rates

Three dimensionless groups characterize the behaviour of the bubbles rising in an infinite liquid: the Eötvös number $Eo = \dfrac{\Delta \rho g d^2}{\sigma}$, the Reynolds number $Re = \dfrac{\rho_1 v_{bb} d}{\mu_f}$ and the Morton number $Mo = \dfrac{g \mu_f^4}{\rho_1 \sigma^3}$ (e.g. Clift et al. 1978), where V_b is the terminal velocity of the bubble, μ_f is the fluid viscosity and other terms are as defined in Table 1. For Lengai carbonatite with $\mu_f \sim 1\,Pa\,s$, and bubble rise velocities of the order of $4\,m\,s^{-1}$ [(Vergniolle and Jaupart 1986; Eq. (C5)], then $Re \sim 10^4$, $Eo \sim 10^5$ and $Mo \sim 0.5$. This places the bubble in the "spherical cap regime" (Clift et al. 1978), where a rising bubble assumes the form of a segment of a sphere, rather than a full sphere. Smaller bubbles in more viscous silicate lavas are characterized by lower Re, and higher Mo and are approximately spherical. The volume of a spherical cap bubble is $\sim 0.8\,y^3$, where y is the cap base radius. Thus, the volume of gas ejected per bubble explosion was 0.8–$2.6\,m^3$, and the time-averaged gas flux was 0.4–$2.4\,m^3\,s^{-1}$. Assuming the gas composition was close to that measured over a similar vent earlier in 1988 ($\sim 50\%$ CO_2, 50% H_2O; Javoy et al. 1989) gives degassing fluxes of 2–$10\,mol/s$ CO_2 and 4–$25\,mol/s$ H_2O at magmatic temperatures ($585\,°C$, Dawson et al. 1990). The low Ne/He ratio and low N_2 concentration (0.011 and $<2\,wt\%$ respectively, Javoy et al. 1989), indicates minimal contamination of the magmatic gas with either air or air-saturated meteoric water.

These gas fluxes are comparable to those measured at Stromboli in 1974 (~ 7–$14\,mol/s$), where individual explosions are considerably larger ($>10^2\,kg$ gas ejected per explosion), but the periodicity is of the order of 10^2–$10^3\,s$ (Chouet et al. 1974; McGetchin et al. 1974; Settle and McGetchin 1980).

3.1.3 Interpretation

The observation of burst periodicity and volumetric eruption rate allow Eqs. (1)–(3) to be solved to determine the characteristic dimensions of the chamber roof. Solutions which satisfy these constraints are summarized in Fig. 3. Calculations assume that the mean bubble size of foam lavas erupted from Hornito 5 ($\sim 1\,mm$ radius) is typical. Equivalent bubble radii at depth were calculated assuming a mixed CO_2–H_2O gas ($1:1$) from a modified Redlich-Kwong equation of state (Ferry and Baumgartner 1987; J.G. Blank, pers. comm.). Envelopes bracket the likely range of solutions: dark stipple indicates solutions for spherical-cap shaped bubbles; fine stipple indicates solutions if, instead, the bubbles are spherical and the gas fluxes consequently higher.

For the conservative case, where the bubbles are assumed to be spherical caps, then the radius of the chamber roof where bubbles are collecting is only of the order of a few metres, and the cross sectional area of the part of the chamber below Centre 11 is $<10^2\,m^2$. The range of possible gas fluxes gives $0.2 > N_2 > 0.02$ [from Eq. (1, 2)], corresponding to the cyclic style of behaviour.

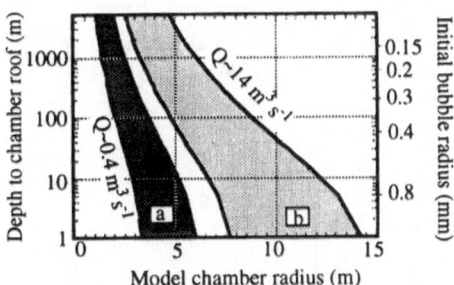

Fig. 3. Summary of constraints on the dimensions of the chamber where gas slugs are generated by foam collapse. Calculations assume that the mean bubble size of the highly vesicular lavas erupted from Hornito 5 (~1 mm radius) is typical. Equivalent bubble radii at depth were calculated assuming a mixed CO_2–H_2O gas (1:1) from a modified Redlich-Kwong equation of state (Ferry and Baumgartner 1987; J.G. Blank, pers. commun.). *Shaded envelopes* bracket the range of solutions, which vary primarily with the estimated eruptive gas flux. The *dark shaded envelope* labelled *a* indicates solutions for spherical-cap bubbles ($Q \sim 0.4$–$2.4\,m^3\,s^{-1}$); *fine stipple* labelled *b* indicates solutions assuming spherical bubbles ($Q \sim 4$–$14\,m^3\,s^{-1}$)

The maximum diameter of the conduit feeding the lava lake can also be estimated since stable gas slugs rising up pipes are generally elongate (Clift et al. 1978), and hence the diameter of an equivalent sphere gives a maximum estimate of the conduit diameter. Since the spherical cap bubbles had volumes of 0.8–$2.6\,m^3$, then for a conservative lava pond depth of 5 m, the maximum diameter of the conduit feeding into the lava pond is of the order of 0.8–$1.3\,m$.

Thus at Centre 11, the presence of a small (<10-m radius) reservoir with an intermittent foam layer along its roof is sufficient to explain the periodicity and volumetric eruption rate of bubbles through the lava lake.

3.2 Centre 5

During the first 1000 min of observation at Hornito 5, the magma level in the hornito fluctuated rapidly, on a time scale as short as minutes, and with an amplitude of up to 1.5 m (Fig. 3). Since Hornito 5 had a cross sectional area of $1.3\,m^2$, these variations correspond to changes in magma volume of <0.005 to $2\,m^3$. Dramatic reductions in magma level invariably correlated with periods of increased eruptive activity from the nearby Centre 5b (Fig. 4), which usually involved the eruption of vesicle-free lava. Magma in Hornito 5 was highly vesicular, with a bulk density as low as $1150\,kg\,m^{-3}$, and a bubble content of up to 47 vol% (Dawson et al. 1990). The simplest explanation of this relationship is that a rising magma level in Hornito 5 corresponded to periods of foam buildup, while sudden drops in the magma level corresponded to episodes of foam collapse (Fig. 5).

For bubble radii of the order of ~0.5 mm (corresponding to a depth of the order of tens of metres, Fig. 3), then from Eq. (5) and using the known properties of natrocarbonatite (Table 1), the critical thickness of foam for

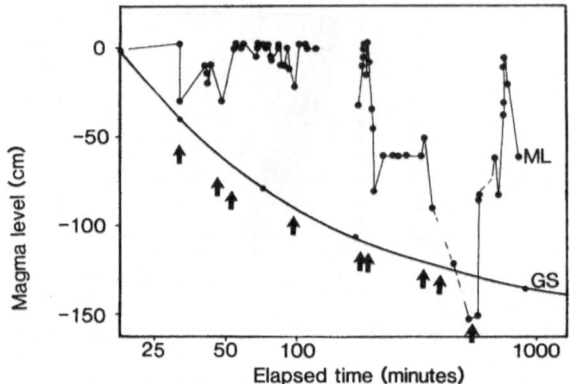

Fig. 4. Summary of measurements of the magma level (ML) in Hornito 5, November 24th. The x-axis shows the time elapsed since 0.5:10 h, when the hornito formed (Pinkerton et al., Chap. 2, this Vol.). The elevation of the top of the hornito is taken as the reference level, relative to which the magma level (*ML*) and ground surface (*GS*) were measured. When magma completely filled Hornito 5, ML = 0. After 1000 min, the hornito was 1.4 m high, so GS = −140 cm. *Solid dots* denote measurements of magma level, or of the height of the hornito. The thick arrows mark periods of violent or enhanced activity at Centre 5b. These invariably correspond to rapid fluctuations of 20–90 cm in the magma level at Hornito 5, indicating that the two centres were connected to the same magma supply system. Over the 1000 min of observation shown, the magma level fluctuated with an amplitude of up to 1.5 m

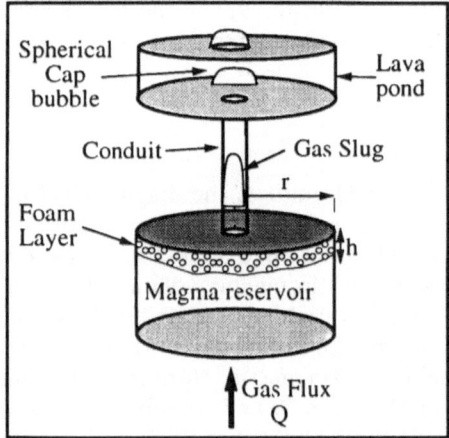

Fig. 5. Cartoon summarizing the foam degassing model (Jaupart and Vergniolle 1988, 1989) as applied to Oldoinyo Lengai. A gas flux Q enters a reservoir of radius r; and bubbles collect below the roof of the reservoir, forming a foam layer of thickness h. Foam collapse leads to the expulsion of gas slugs out of a smaller conduit. As these slugs flow through the lava pond, they adopt the form of a spherical cap, before bursting as they reach the surface. The foam layer is everywhere <6 cm thick, and is trapped at the top of a chamber with a diameter of <20 m, with periodic collapse and explosion of the layer every 1–2 s

collapse is ~6 cm. The $1-2\,m^3$ volume fluctuations observed at Hornito 5 could clearly be accommodated by intermittent collapse of a foam layer accumulating below a roof area of only a few tens of square metres. This is of the same order as the earlier estimates of the dimensions of the conduit supplying Centre 11, which seems reasonable. Similar mechanisms have been proposed to explain periodic magma level changes at Kilauea (Tilling 1987; Vergniolle and Jaupart 1990).

4 Discussion and Conclusions

The bubble explosions in the lava lake at Centre 11, and the fluctuations in magma level at Centre 5 are both consistent with a model of periodic foam accumulation and collapse at the roof of a shallow magma reservoir. The order of magnitude constraints on the dimensions of the reservoir below Centre 11 require a small chamber, with a radius of <10 m to exist below the active November 1988 vent. A similarly sized (~10 m scale) reservoir is required below Centre 5, to accommodate the $1-2\,m^3$ volume fluctuations if these were generated by foam collapse. These constraints suggest that magma systems immediately below Centres 5 and 11 were not directly connected to the same sub-chamber (cf. Fig. 1).

Instead, a hierarchy of conduits from the final vents (<1–2 m width) and sub-chambers (<5–10 m radius) were probably sourced from a deeper plumbing system. If there is a stored magma volume approaching $10^6-10^7\,m^3$ below Oldoinyo Lengai, as suggested by modelling of the radioactive disequilibria of the 1988 lavas (Pyle et al. 1991; Pyle, Chap. 9, this Vol.), then most of this volume cannot be accommodated in these very shallow conduits and sub-chambers, and we predict that there must be a deeper and larger reservoir, perhaps with a radius approaching the dimensions of the 1966 pit crater (~100 m).

Acknowledgements. We thank the Roayal Society for financial support, and Celia Nyamweru, the Peterson Brothers and Dorobo Safaris for field assistance. Comments by Gail Mahood and two anonymous reviewers are gratefully acknowledged. DMP was supported by a St. Chatharine's College, Cambridge Research Fellowship and a Visiting Associateship from the California Institute of Technology, and thanks Ed Stolper and the third floor residents for sharing their facilities and hospitality. Jen Blank is acknowledged for the MRK-EOS program. Department of Earth Sciences contribution number 2531.

References

Blackburn EA, Wilson L, Sparks RSJ (1976) Mechanisms and dynamics of strombolian activity. J Geol Soc Lond 132:429–440

Chouet B, Hamisevicz N, McGetchin TR (1974) Photoballistics of volcanic jet activity at Stromboli, Italy. J Geophys Res 79:4961–4976

Clift R, Grace JR, Weber ME (1978) Bubbles, drops and particles. Academic Press, New York, pp 1–380

Dawson JB (1962) The geology of Oldoinyo Lengai. Bull Volcanol 24:349–387
Dawson JB (1989) Sodium carbonatite extrusions from Oldoinyo Lengai, Tanzania: implications for carbonatite complex genesis. In: Bell K (ed) Carbonatities – genesis and evolution. Unwin and Hyman, London, pp 255–277
Dawson JB, Bowden P, Clark GC (1968) Activity of the carbonatite volcano, Oldoinyo Lengai. Geol Rundsch 57:865–879
Dawson JB, Pinkerton H, Norton GE, Pyle DM (1990) Physicochemical properties of alkali carbonatite lavas: data from the 1988 eruption of Oldoinyo Lengai, Tanzania. Geology 18:260–263
Ferry JM, Baumgartner L (1987) Thermodynamic models of molecular fluids at the elevated pressures and temperatures of crustal metamorphism. Rev Mineral 17:323–365
Hobley CW (1918) A volcanic eruption in east Africa. J East Afr Uganda Nat Hist Soc 5:339–343
Janz GJ (1988) Thermodynamic and transport properties for molten salts: correlation equations for critically evaluated density, surface tension, electrical conductance and viscosity data. J Phys Chem Ref Data 17 Suppl 2:1–309
Jaupart C, Vergniolle S (1988) Laboratory models of Hawaiian and Strombolian eruptions. Nature 331:58–60
Jaupart C, Vergniolle S (1989) The generation and collapse of a foam layer at the roof of a basaltic magma chamber. J Fluid Mechanics 203:347–380
Javoy M, Pineau F, Staudacher T, Cheminee JL, Krafft M (1989) Mantle volatiles sampled from a continental rift: the 1988 eruption of Oldoinyo Lengai (Tanzania). Terra Abstr 1:324
Keller J, Krafft M (1990) Effusive natrocarbonatite activity of Oldoinyo Lengai, June 1988. Bull Volcanol 52:629–645
Krafft M, Keller J (1989) Temperature measurements in carbonatite lava lakes and flows from Oldoinyo Lengai, Tanzania. Science 245:168–170
McGetchin TR, Settle M, Chouet B (1974) Cinder cone growth modelled after Northeast crater, Mount Etna, Sicily. J Geophys Res 79:3257–3272
Nyamweru C (1988) Activity of Oldoinyo Lengai volcano, Tanzania 1983–1987. J Afr Earth Sci 7:603–610
Nyamweru C (1989) Report of activity in the northerm crater of Oldoinyo Lengai, 24th June to 1st July 1988. J East Afr Nat Hist Soc Natl Mus 79:1–15
Pyle DM, Dawson JB, Ivanovich M (1991) Short-lived decay series disequilibria in the natrocarbonatite lavas of Oldoinyo Lengai, Tanzania: constraints on the timing of magma genesis. Earth Planet Sci Lett 105:378–396
Settle M, McGetchin TR (1980) Statistical analysis of persistent explosive activity at Stromboli, 1971: implications for eruption prediction J Volcanol Geotherm Res 8:45–58
Sparks RSJ (1978) The dynamics of bubble formation and growth in magmas: a review and analysis. J Volcanol Geotherm Res 3:1–37
Tilling RI (1987) Fluctuations in surface height of active lava lakes during 1972–1974 Mauna Ulu eruption, Kilauea volcano, Hawaii. J Geophys Res 92:13721–13730
Vergniolle S, Jaupart C (1986) Separated two-phase flow and basaltic eruptions. J Geophys Res 91:12842–12860
Vergniolle S, Jaupart C (1990) Dynamics of degassing at Kilaeua volcano, Hawaii. J Geophys Res 95:2793–2809
Wilson L (1972) Explosive volcanic eruptions – II. The atmospheric trajectories of pyroclasts. Geophys J R Astr Soc 30:381–392
Wilson L (1980) Relationships between pressure, volatile content and ejecta velocity in three types of volcanic explosion. J Volcanol Geotherm Res 8:297–313

Petrology and Geochemistry
of Oldoinyo Lengai Lavas Extruded in November 1988:
Magma Source, Ascent and Crystallization

J.B. Dawson[1], H. Pinkerton[2], G.E. Norton[3], D.M. Pyle[4], P. Browning[5],
D. Jackson[6], and A.E. Fallick[7]

Abstract

Lavas erupted from Oldoinyo Lengai in November, 1988, carry phenocrysts of the alkali carbonates nyerereite and gregoryite, with inclusions of apatite. They are set in a matrix of microphenocrysts of nyerereite and gregoryite wih tiny grains of $(Mn, Fe)S$, MnFe spinel, sodic sylvite, Na- and Si-rich apatite, spurrite, and intergrowths of sylvite, fluorite and a phase similar to nyerereite. The matrix also contains an unidentified complex carbonate? of Ca, Ba, Sr, K and Na. All phases represent solid solutions (attributed to the high-temperature crystallization of the chemically complex dominantly carbonate liquid) that are very rare or not previously reported.

Chemically, the lavas are very similar to those erupted in 1960, containing high amounts of Na_2O, K_2O, CaO, CO_2 and lesser, but nonetheless significant, amounts of BaO, SrO, Cl, F, P_2O_5 and SO_3. Trace element concentrations and REE patterns indicate that the carbonatite is highly fractionated. Compositional variations arise mainly between phenocryst-rich and aphyric varieties, with the aphyric variety being richer in K, Ba, Cl, F and Rb. More subtle variations in Fe, Mn, Pb and Zn may reflect groundmass sulphide sedimentation. The isotope ratios of carbon, oxygen and sulphur are similar to those in other primary carbonatites. Lead and thorium isotope data, combined with previously published Sr and Nd isotope data, indicate that the carbonate magma originates from an upper mantle source with the same isotopic characteristics as mantle sources for ocean island basalts. The isotopic data provide no evidence for interaction between crustal material and the carbonatite or its parental magma during ascent from the mantle source.

1 Introduction

Oldoinyo Lengai is the world's only active carbonatite volcano. The alkali-rich composition of the carbonatite lavas was first recognized for lavas extruded in 1960 (Dawson 1962a,b). Lavas continued to be extruded until late July 1966; the

[1] Grant Institute of Geology, University of Edinburgh, West Mains Road, Edinburgh EH9 3JW, UK
[2] Environmental Science Division, University of Lancaster, Lancaster LA1 4YQ, UK
[3] British Geological Survey, Keyworth, Nottingham NG12 5GG, UK
[4] Department of Earth Sciences, University of Cambridge, Downing Street, Cambridge CD2 3EQ, UK
[5] Department of Geology, University of Bristol, Bristol BS8 1RJ, UK
[6] Department of Earth Sciences, Open University, Milton Keynes MK7 6AA, UK
[7] Scottish Universities Research and Reactor Centre, East Kilbride G75 0QU, UK

compositions of the extrusions throughout this entire period is not known for certain, though samples collected in 1961 (Du Bois et al. 1963) and 1963 (Peterson 1990) were compositionally similar to the 1960 lavas. Ash from violent eruptions in August–October 1966 and July 1967 infilled the active crater and the ash cone collapsed to form a precipitously sided pit crater (Dawson et al., Chap. 1, this Vol.). The volcano then became dormant, and remained so until January 1983 when minor ash eruptions heralded a new phase of activity. Small-volume lavas extruded onto the floor of the pit crater in April 1983 and slowly began to fill up the crater (Nyamweru 1988). In June 1988, the volcano was visited by an expedition led by M. Krafft and J. Keller. The results of this visit are given by Krafft and Keller (1989) and Keller and Krafft (1989, 1990). In brief, the activity, flow morphology and lava chemistry were similar to those in 1960; lava temperatures in flows were 491–519 °C and 544 °C in a lava lake.

The first four authors, with C. Nyamweru, visited the volcano in the period 21–25 November 1988. Since the Krafft expedition in June, the crater floor had risen because of flows from several new cones and craters; in particular, the Southern Depression, a topographically low area separated from the active part of the crater by a tuff ridge (the Saddle – Fig. 1), was being filled by lava flows from a major vent – Centre 11. Flows continued from this and other centres during the period of observation during which we made observations on flow morphology and the physical properties (temperature and apparent viscosity) of the lavas (Dawson et al. 1990). In summary, the apparent viscosity varied from 0.3 to 120 Pa s, with highest values being obtained on frothy, phenocryst-rich

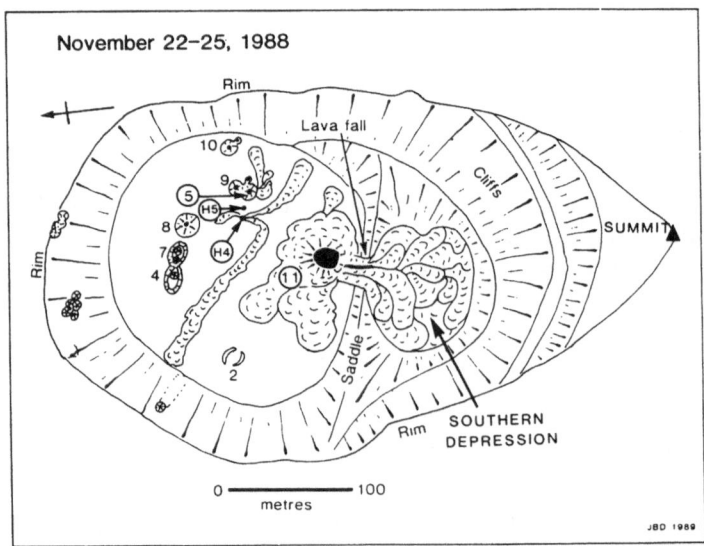

Fig. 1. Plan of the active northern crater of Oldoinyo Lengai, showing flows (*ornamented*) extruded in the period 22–25 November 1988. *Circled numbers* and *circled H numbers* refer to extrusion vents and hornitoes, respectively

lava and the lowest on gas-poor, phenocryst-poor lava. The extrusion temperatures were 585 ± 10 °C, which is higher than temperatures of 491–544 °C reported by Krafft and Keller (1989). Nonetheless, the temperatures and viscosities of these alkali carbonatites are the lowest recorded for terrestrial magmas.

This chapter is concerned with the petrology and geochemistry of lavas that we collected in November 1988. They were extruded from the centres active at that time – Centres 5 and 11 and Hornitos 4 and 5 (Fig. 1). Of particular relevance is an observation that Centre 5 and the adjacent Hornito 5 were fed from the same magma pool; lava extrusion from Centre 5 was sporadic, and the extrusion of the gas-free, very mobile, aphyric lava from the base of the Centre 5 cone coincided with falls in the level of the frothing, phenocryst-rich lava in Hornito 5; conversely, cessations in lava extrusion from Centre 5 coincided with rises in the level of the lava pool in Hornito 5.

2 Petrography

The lavas consist of phenocrysts of the alkali carbonates nyerereite (Na, K)$_2$ Ca(CO$_3$)$_2$ and gregoryite (Na$_{0.78}$K$_{0.05}$)$_2$Ca$_{0.17}$(CO$_3$) with microphenocrysts of nyerereite and gregoryite in a fine-grained matrix. The compositions of both phases given above are simplifications, as there are substitutions of Sr and Ba for Ca, and of SO$_3$, F, Cl and P$_2$O$_5$ for CO$_3$.

The lavas have a variety of textures reflecting their residence time and site in the vents and their mode of extrusion. At one end of the spectrum is phenocryst-poor, gas-poor lava from highly mobile, 1- to 5-cm-thick lava flowing from fissures on the lower flanks of Centres 5 and 11 (see Fig. 4, Pinkerton et al., Chap. 2, this Vol.); these flows have a pronounced trachytic texture due to alignment of nyerereite microphenocrysts. In contrast, highly vesicular, frothy lavas from Hornito 5 and other lava pools have a high proportion of nyerereite

Table 1. Modes of the two most texturally extreme lavas (in vol%)

	BD4155A[b]	BD4155B[b]	BD4169
	Highly mobile, gas-free		Viscous, vesicular
Gregoryite phenocrysts	6.5	0	30.4
Nyerereite phenocrysts	9.1	0	31.5
Nyerereite microphenocrysts	13.1	16.2	0
Void (vesicle)	3.3	0	13.9
Groundmass[a]	67.9	83.8	24.1
Maximum size of phenocrysts	1 mm		2.3 mm

[a] Includes opaques.
[b] Specimen is banded, with A = porphyritic, and B = aphyric bands.

and gregoryite phenocrysts (Table 1), consistent with flotation in the convecting lava pools. The two texturally extreme types are interpreted as representing complementary fractions of the carbonate magmas – the gas-free liquid fraction being tapped from a deeper level in the lava pools, the other a phenocryst-rich lava-pool surface froth. Many flows represent mixtures. The most extreme type of the gas-free variety occurs as minor squeeze-ups or driblets on the surface of flows or at the toes of mobile flows; these are essentially phenocryst-free. Whereas the driblets contain the highest concentrations of the less dense groundmass phases, low Fe, Mn and Pb contents suggest that most of the denser groundmass sulphides had sedimented out closer to the lava pools before breakout of the driblets from their parent flows.

The clear, stumpy, prismatic phenocrysts of nyerereite are up to 3 mm; prism sections are euhedral and show third-order red-green birefringence, whereas rounded basal sections show polysynthetic twinning and first-order grey-white birefringence. The only inclusions are of stumpy apatite grains up to 20 μm long. Acicular nyerereite microphenocrysts are up to 0.7 mm long, and often show the "swallow-tail" bifurcation typical of quench phases. The typically rounded gregoryite phenocrysts are up to 3 mm, have marked cleavage and have a microperthitic texture. The cores are light brown in plane-polarized light, contain inclusions of apatite and (?exsolved) lamellae of nyerereite, and are turbid due to numerous minute inclusions of a weakly translucent phase; the cores are zoned, differing zones having different inclusion concentrations, and are overgrown by clear, inclusion-free rims. Gregoryite microphenocrysts are clear and unzoned.

Phases in the groundmass are rarely >20 μm in size; as a result, some identifications and qualitative analyses have been made by scanning electron microscopy (with an attached energy dispersive analytical system – EDS) and back-scattered electron imaging (BSEI). The opaque sulphide grains (up to 20 μm) vary from equant octahedra to amoeboid blebs and globular, coalescing clusters; many grains are spongiform, and BSE images show that some have compositionally different rims. The grains of opaque oxide are small (<5 μm) and some have compositionally different rims. The rounded sylvite grains, first confirmed as a primary magmatic phase by Keller and Krafft (1989) and Peterson (1990), are up to 15 μm. The groundmass apatite grains, up to 20 μm, are chemically zoned, with BSE imaging showing slightly brighter (denser) 3–4 μm rims that reflect slightly higher concentrations of REE; in this respect they differ from the unzoned apatites included in the nyerereite and gregoryite phenocrysts. A Ca-silicate containing detectable amounts of REE zones towards Fe-rich rims in some grains. Elliptical, well-bounded intergrowths up to to 30 μm (Fig. 2) consist of acicular sylvite and fluorite intergrown with a phase similar in composition to nyerereite, but containing more CaO relative to K_2O and containing more Cl and S than nyerereite; fluorite has been found only in the intergrowths. Another phase – a Ca- and Sr-bearing high-Ba carbonate (referred to hereafter as phase X) – occurs most commonly as a partial overgrowth on nyerereite microphenocrysts and protruding into the interstices between other groundmass phases (Fig. 2). Other Ba-rich phases are barite and witherite that occur in rare 10 μm grains.

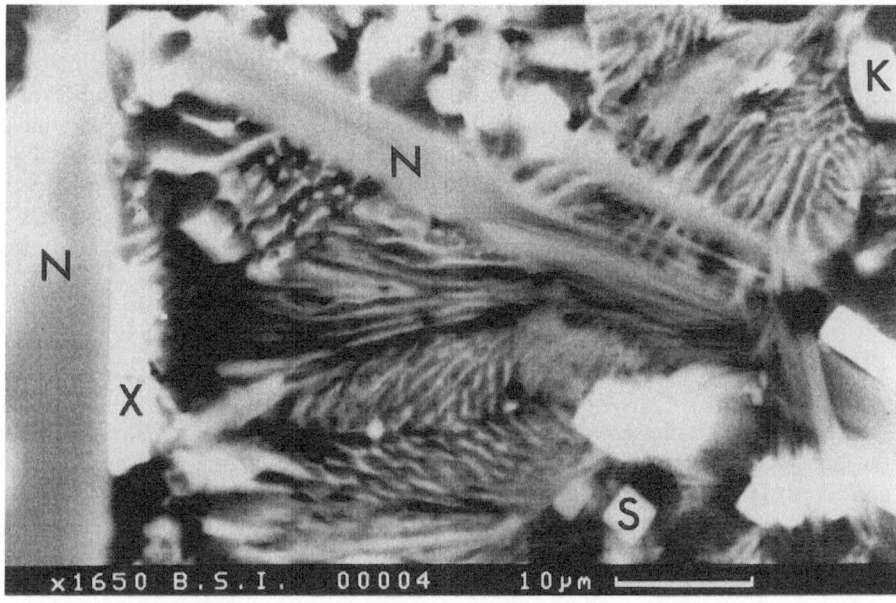

Fig. 2. BSE image of the matrix in BD4155 showing patches of phase X (X) alongside a bladed microphenocryst of nyerereite (N), rounded grains of sylvite (K), equant grains of sulphide (S) and intergrowths of acicular sylvite, fluorite and "nyerereite"

The petrography indicates two stages of cooling. Phenocrysts of gregoryite and nyerereite were the first to crystallize, but the gregoryite zoning suggests that it had a more complex growth history; however, both phases continued to precipitate in the matrix. Initial apatite crystallization overlapped with that of the phenocrysts and it continued to precipitate throughout cooling, becoming more REE-enriched in the groundmass. Rapid cooling upon extrusion resulted in crystallization of the residual liquid in which Fe, Mn, S, REE, Ba and the halogens had been relatively concentrated.

3 Mineral Chemistry

Mineral analyses were made by WDS on a Cameca microprobe using PAP correction in the Department of Geology and Geophysics at the University of Edinburgh. Operating conditions were 20 nA and 20 kV. The beam size was 2–3 μm and hence it was inevitable that, in the analysis of some of the smaller grains and particularly the rims, there was some overlap with other matrix phases. Phases in the matrix were checked for homogeneity by BSE imaging, and EDS spectra were scrutinized before WDS analysis. Absence of an element in the tables of analyses indicates its absence on the EDS spectrum after a 200-s count, though F and CO_2 are not detectable. Standards for the WDS analyses are Na – Jadeite, Mg – periclase, Al – corundum, Si and Ca – wollastonite, K – orthoclase, Ti – rutile, Fe and Mn – metal, Ba and S – barite, Sr – celestine, REE – synthetic glasses. In studies on the matrix, attention has been concentrated on one particular specimen, BD4155, an aphyric lava collected from a minor overspill from the lava channel below Centre 11; this has proved to have relatively high concentrations of groundmass sulphides, possibly because it had travelled only 10–15 m from Centre 11 before the early consolidation preempted sulphide sedimentation. When analyzing the sulphides, Na, K and Ca were included in the routine to monitor possible overlap with other groundmass phases.

3.1 Nyerereite and Gregoryite

Analyses of these phases, with very minor exceptions, are within the ranges of the analyses that have been published recently by Peterson (1990) and Keller and Krafft (1990) in studies on the 1963 and June 1988 lavas, respectively; hence it is unnecessary to present further analyses here.

McKie and Frankis (1977) state that F is present in nyerereite in a 1960 lava, based on a wet-chemical analysis. Recent developments in analyzing crystals has made accurate analysis of F feasible with the electron microprobe and, during analyses made at the University of Sheffield using a Si/W multilayer crystal, no counts above background for F were detected over repeated 60-s counting periods. We deduce that there is no F in the nyerereite crystals in our lavas.

3.2 Sulphides

Alabandite (MnS) was reported in unspecified lavas by Gittins and McKie (1980). Peterson (1990) provides an analysis of sulphide in the 1963 lava that consists of 66% alabandite, 30% pyrite and 4% pyrrhotite, and Keller and Krafft (1990) report the sulphide in the June 1988 lavas to be Fe-alabandite of composition $Mn_{0.7}Fe_{0.3}S$. The sulphides in specimen BD4155, the one chosen for most intensive study of the groundmass phases, are more varied than in these other reports. The most common sulphide is compositionally most similar to that reported by Keller and Krafft; it is an alabandite-pyrrhotite solid solution having the formula $Mn_{0.74}Fe_{0.26}S$. Additionally, it contains small amounts of Zn, Pb and Ti (Table 2, analysis 1); Cu, Co, Ni and As were sought but not found.

This is a very unusual phase, as terrestrial alabandites contain no Fe and, conversely, pyrrhotites contain no Mn. MnS is known to have extensive miscibility with FeS at high temperatures; one-atmosphere experiments show increasing FeS solid solution with increasing temperature, reaching around 80% FeS molecule at 1150 °C (Skinner and Luce 1971). At the eruption temperature of the carbonatite lavas (585 °C), the maximum predicted solid solution is ~55% FeS molecule. Although Fe-alabandite is found in meteorites, it is rare in terrestrial rocks, but has been recorded (optical determinations) in Mn-rich ores formed by high-temperature contact metamorphism of inclusions in basalt near

Table 2. Compositions of sulphides in lava BD4155

	1	2	3	4	5
Fe	16.2	25.2	20.7	46.1	44.8
Mn	44.3	14.5	30.2	1.66	–
Zn	0.14	19.9	8.49	0.34	–
Pb	0.21	0.18	1.28	0.22	–
Ti	0.03	0.10	0.07	0	–
S	35.2	31.8	32.9	37.6	37.5
Na_2O	0.63	2.61	1.66	0.20	0.06
K_2O	0.11	0.50	0.51	11.8[a]	16.3[a]
CaO	0.51	0.21	0.95	0.16	–
CO_2 calc	0.99	2.74	2.17	0.26	–
Sum	98.32	98.74	98.93	98.34	98.66

[a] K given as element concentration.

Fe	0.290	0.451	0.371	0.831
Mn	0.806	0.264	0.550	0.030
Zn	0.002	0.304	0.130	0.005
Pb	0.001	0.001	0.006	0.001
Ti	0.000	0	0	0
K	–	–	–	0.302
Σ	1.099	1.020	1.057	1.169
S	1.097	0.993	1.028	1.173

1. Mean of four 10-µm grains of "normal" composite $(Mn_{0.74} Fe_{0.26})S$.
2. 2-µm rim on "normal" grain $(Mn_{0.28} Fe_{0.44} Zn_{0.30})S$.
3. Zn- and Pb-bearing grain $(Mn_{0.47} Fe_{0.35} Zn_{0.12} Pb_{0.06})S$.
4. K-Fe sulphide grain $K(Fe, Mn)_3 S_4$.
5. Rasvumite $(K Fe_2 S_3)$, Khibina, USSR (Czamanske et al. 1978).

Kassel, Germany, and in the Fohberg wollastonite phonolite of the Kaiserstubl carbonatite volcano (Ramdohr 1957).

Rarely the more common sulphide grains have thin 5 µm rims of a high-Zn (19.9 wt%) sulphide (Table 2, analysis 2); in this analysis, relatively high amounts of Na, Ca and K (included in the analytical programme for monitoring purposes) indicate some overlap with the surrounding groundmass. There are also rare discrete grains of a Zn-Pb-bearing Mn-Fe sulphide (Zn 8.49 wt%, Pb 1.28 wt%) (Table 2, analysis 3) and rare 5 µm globules of pure PbS. One grain of a K-bearing sulphide was found (Table 2, analysis 4); the grain was homogeneous and inclusion-free, and low amounts of Na and Ca, with absence of Cl, in the analysis indicate minimal overlap with surrounding carbonates or sylvite; hence the 11.8 wt% K is believed to be real. Like the other sulphides, it contains Mn and small amounts of Zn and Pb. Its structural formula approximates $K(Fe, Mn, Zn, Pb)_3S_4$. The main K-Fe sulphides reported in the literature are djerfisherite, rasvumite and bartonite (Czamanske et al. 1978, 1979; Clarke 1979). Djerfisherite and bartonite contain Cu, Ni and Cl and have K:Fe:S ratios differing from those in the Lengai sulphide. Rasvumite (KFe_2S_4), although lacking the Cu, Ni and Cl of the other phases, is unlike the Lengai sulphide in lacking Mn, Zn and Pb and having a different structural formula. In short, the Lengai K-sulphide is unlike other reported K-sulphides.

3.3 Oxide

Shive et al. (1990) report a spinel of the magnetite-jacobsite series ($Fe_{2.13}Mn_{0.54}$ $Mg_{0.28}Zn_{0.01}Ti_{0.02})O_4$ in lavas collected on Oldoinyo Lengai in November 1987 and June 1988. Opaque 2–3-μm grains of a Fe-Mn oxide occur in the groundmass of our lavas, but are less abundant than sulphide. Their size causes analytical problems due to overlap with other phases, resulting in low totals. A typical analysis gives SiO_2 1.29, TiO_2 0.72, FeO 73.3, MnO 10.3, TiO_2 0.73, MgO 1.93, CaO 1.29, BaO 0.21, Na_2O 0.93, K_2O 0.33, sum 90.46 wt%. Assuming a spinel structure, the structural formula for this (less alkalies, CaO and SiO_2) is ($Fe_{2.34}Mn_{0.34}Ti_{0.20}Mg_{0.11}$) O_4. There is complete miscibility between Fe_3O_4 and Mn_3O_4 above 1000 °C (Mason 1947) but there are no data for around 600 °C. With sulphide, this phase is the major site for Fe and Mn in the lavas; Mn partitions preferentially into the sulphide and Fe into the oxide.

3.4 Ca-Silicate

In addition to Ca and Si, the Ca-silicate contains appreciable amounts of Fe, Mn, Ba, Na, K, La and Ce (Table 3). A rim on one grain is enhanced in Fe, Mn, Mg and Ba relative to the other grains (Table 3, analysis 2). If the low sums

Table 3. Analyses of Ca-silicates in lavas BD4155 and 4156

	1	2	3	1		2	
SiO_2	27.34	23.15	27.25	Si 2.004		1.900	
FeO	1.53	21.18	0.10	Fe 0.092		1.451	
MnO	0.50	3.17	0.02	Mn 0.031		0.232	
MgO	0.18	0.80	0.19	Mg 0.017		0.019	
CaO	51.62	37.46	62.78	Ca 4.018		3.302	
BaO	0.13	0.57	n.a	Ba 0.004	4.697	0.019	5.435
Na_2O	2.96	2.08	0.03	Na 0.422		0.296	
K_2O	0.68	0.59	–	K 0.061		0.059	
La_2O_3	0.84	0.83	n.a.	La 0.017		0.019	
Ce_2O_3	1.42	1.42	n.a.	Ce 0.035		0.038	
CO_2	12.80[a]	8.75[a]	8.97	C 1.275		0.982	
Sum	100.00	100.00	99.34				

[a] CO_2 calculated by differences.
1. Mean of eight grains of Ca-silicate in 4155 and 4156.
2. Mean of two analyses on a Fe-rich rim on normal grain in 4155.
3. Paraspurrite, Darwin, California (Colville and Colville 1977).

are assumed to be due to CO_2 in the lattice, the mineral is most similar to the rare group of calcic mixed silicate-carbonate phases that includes spurrite and tilleyite. The ratio between Si and the other elements (excluding assumed C) of approximately 5:2 fits with either spurrite or tilleyite, but the amount of assumed CO_2 is more consistent with spurrite (Table 3, analysis 3). We tentatively identify the phase as an alkali-REE-Fe variant of spurrite, but X-ray confirmation is required. This is the only silicate phase that we have identified. Peterson (1990) found rare grains of Mn-monticellite in the 1963 lavas, and Keller and Krafft report a phase of larnite-spurrite affinities in the June 1988 lavas. Dawson et al. (1992) found a sodic calcium disilicate (possibly larnite – Na_2O 1–2 wt%) and strontian rankinite in combeite reaction coronas around wollastonite and clinopyroxene crystals in lapilli from the 1966 ash eruption of Oldoinyo Lengai.

3.5 Apatite

The small phosphate grains are unusual in that they are particularly high in alkalies (Table 4), even exceeding the high values of 3.09 wt% Na_2O found in apatites in South Greenland peralkaline pegmatites (Rønsbo 1989). The Lengai apatites are also high in SiO_2 and halogens, particularly F. These apart, the oxide values fall within the ranges given by Hogarth (1989) in a review of carbonatite apatites. As may be expected, they contain significant SrO and high light REE. No matter what coupled substitutions involving Ca, P, Si, Na and REE are used (see Hogarth 1989 and Rønsbo 1989), the high alkali and relatively

Table 4. Analyses of apatite

	1	2		1	2
P_2O_5	29.80	30.82	P	2.372 } 2.595	2.387 } 2.617
SiO_2	2.47	2.42	Si	0.223	0.230
CaO	42.81	43.65	Ca	4.214	4.279
SrO	3.50	3.63	Sr	0.191	0.193
BaO	1.46	1.45	Ba	0.054	0.052
La_2O_3	1.65	1.85	La	0.057	0.062
Ce_2O_3	2.60	2.64	Ce	0.090 } 6.092	0.088 } 6.281
Pr_2O_3	0.15	0.07	Pr	0.005	0.002
Nd_2O_3	0.39	0.44	Nd	0.013	0.014
Na_2O	7.85	7.44	Na	1.432	1.320
K_2O	3.00	2.32	K	0.360	0.271
F	4.48	5.27	F	1.339 } 1.461	1.524 } 1.619
Cl	0.77	0.61	Cl	0.122	0.095
Sum	100.93	102.61	O 12		
Less O≡F,Cl	2.06	2.36			
	98.87	100.25			

1, 2. Core and rim of groundmass grain, respectively.

Table 5. Analyses of three patches of phase X

n^a =	1 2	2 3	3 3
CaO	30.77	18.99	11.31
MgO	0.75	3.18	3.72
SrO	5.56	6.96	7.38
BaO	18.58	34.40	44.09
Na_2O	2.79	2.52	2.83
K_2O	8.42	3.13	3.17
La_2O_3	0.26	0.56	1.11
F	1.64	0.23	1.07
Cl	3.63	0.22	1.01
Sum	72.40	70.49	75.79

[a] Number of analyses per patch.

low P_2O_5 contents cause major departures from the ideal formula. Whilst it is possible that these aberrations may be due to the presence of microinclusions (a feature of apatites in general), other high-alkali phosphates (P_2O_5 19–24, SiO_2 2, CaO 19–22, Na_2O 26–31, K_2O 4–6, Cl 0.2–0.4 wt%) are known to occur in lapilli from the 1966 ash eruption (Dawson et al. 1992).

3.6 Phase "X"

Phase X is an enigma. It occurs as elongate or irregular grains or patches that, under BSE imaging sufficient to resolve other groundmass intergrowths, appear to be homogeneous. If it is a mixture of different phases, the intergrowths are at the sub-micron level or consist of phases of identical back-scattering coefficients. It is the principal site for Ba within the groundmass, and consistently contains MgO, CaO, SrO, Na_2O, K_2O, La_2O_3, F and Cl (Table 5); it is the most magnesian of all the phases. Within individual patches, the composition is consistent but there is considerable variation between patches, particularly with respect to CaO, BaO, K_2O and Cl (Table 5); Ba, Mg, Sr and La collectively vary inversely with Ca. K varies with Cl, perhaps suggesting a little overlap with sylvite during analysis, though there is a considerable excess of K over that required to combine with Cl to form sylvite. Low analytical sums suggest the presence of an unanalyzed element, most plausibly C in the present paragenesis; hence we tentatively identify phase X as a largely carbonate solid solution expressible in terms of variable proportions of the major end-members calcite-witherite-strontianite-$(Na,K)_2CO_3$.

3.7 Sylvite

The small sylvite grains have proved difficult to analyze because of a poor polish due to their intrinsic softness; this has resulted in low analytical totals. In

Table 6. Storage sites of minor and trace elements

Fe	SULPHIDE, SPINEL
Mn	SULPHIDE, Spinel
Mg	PHASE X, Spinel, Sellaite
P	APATITE, gregoryite
S	SULPHIDE, gregoryite, nyerereite
Ba	PHASE X, barite, witherite, nyerereite
Sr	Nyerereite, phase X, apatite
Cl	SYLVITE, gregoryite, phase X
F	FLUORITE, apatite, gregoryite
REE	APATITE, Ca silicate, phase X

Phases in UPPER CASE are the major storage site(s).

addition to K and Cl, other elements found during analysis are Na, Ca, Ba, F, S and P. Of these, only Na varies positively with K and Cl and, like K and Cl, reaches its highest concentrations in analyses with highest totals. Hence we believe the Na to be sited in the sylvite, unlike the other elements, which we attribute to peripheral fluorescence.

One result of the electron-microprobe analyses is that we can now clearly identify the mineralogical siting of the minor and some trace elements; these are summarized in Table 6.

4 Geochemistry

Major and trace element analyses of lavas collected during the expedition, together with a sample collected by C. Nyamweru during the June eruption, are given in Tables 7 and 8; analyses of two lavas erupted in 1960 are given for comparison.

In general terms, the 1988 lavas erupted in June (reported by Keller and Krafft 1990) and November (reported here) are similar to those erupted in 1960, containing high amounts of Na_2O, K_2O, CaO and CO_2, and with lesser, but nonetheless quite high, amounts of BaO, SrO, P_2O_5, SO_3, Cl and F. Other oxides are in amounts <1 wt%. The only minor oxide of note is MnO, which is generally in concentrations greater than FeO and MgO.

Despite the overall similarities in the bulk compositions of the lavas, there is a range in chemistry which can be linked to textural type (Dawson et al. 1990; Keller and Krafft 1990). The extremes in composition are BD4167, mobile, gas-free, aphyric lava from Centre 5, and BD4169, viscous, vesicular, phenocryst-rich lava from the nearby (and apparently interconnected) Hornito 5 lava pool. The other lavas fall within the compositional ranges delimited by these two specimens. The relatively matrix-rich BD4167 is richer in BaO, K_2O, Cl, F, V, Cu and Rb, but poorer in CaO, CO_2 and P_2O_5. Ranges in the other major and trace elements are not significant, but the differences noted give rise to lower

Table 7. Analyses of 1988 carbonatite lavas

	1	2	3	4	5	6	7	8	A	B
SiO_2	0.29	0.14	0.17	0.19	0.22	0.23	0.18	0.18	Tr	Tr
TiO_2	0.02	0.01	<0.01	0.01	0.01	0.02	<0.01	0.01	0.11	0.08
Al_2O_3	0.10	0.10	0.10	0.10	0.12	0.10	0.10	0.10	0.09	0.09
Fe_2O_3	0.50	0	0.13	0.20	0.21	0.36	0.21	0.16	0.28[a]	0.35[a]
FeO	0.24	0.10	0.06	0.14	0.08	0.10	0.12	0.10	0.0	0.0
MnO	0.61	0.24	0.24	0.42	0.31	0.38	0.45	0.34	0.04	0.04
MgO	0.42	0.46	0.27	0.28	0.27	0.26	0.22	0.29	0.53	0.58
CaO	13.36	13.04	14.15	13.81	13.52	14.18	14.63	13.93	13.90	13.96
BaO	1.68	1.92	1.30	1.39	1.47	1.29	1.05	1.36	1.04	1.14
SrO	1.74	1.79	1.68	1.69	1.70	1.69	1.63	1.71	1.53	1.31
Na_2O	31.06	32.58	32.03	32.18	31.92	32.45	32.72	30.74	32.22	32.35
K_2O	8.44	8.64	7.75	8.11	8.22	7.78	7.27	7.93	8.27	7.17
H_2O^+	0.05	0.23	0.03	0.03	0.0	0.18	0.22	0.11	0	0
P_2O_5	0.73	0.63	0.91	0.86	0.86	0.99	1.02	0.89	0.90	0.98
SO_3	3.12	3.07	2.92	3.17	3.10	2.88	3.18	3.07	2.18	2.37
CO_2	32.7	29.2	35.0	34.7	34.3	35.0	34.3	35.2	34.65	35.29
Cl	3.85	4.40	2.25	2.50	2.60	1.60	1.80	2.65	4.21	2.87
F	3.55	5.25	2.50	2.20	2.10	1.65	1.35	2.75	2.93	3.19
Sum	102.47	101.80	101.48	101.98	100.97	101.34	100.45	101.52	102.88	101.77
Less										
O=Cl and F	2.35	2.03	1.55	1.48	1.46	1.05	0.97	1.75	2.18	1.48
Total	100.12	99.77	99.93	100.50	99.51	100.29	99.48	99.77	100.70	100.29

[a] Total iron expressed as Fe_2O_3.
Analyst: A. Saxby   Analyses by XRF (except CO_2, FeO, H_2O, Cl and F – wet methods)
1.   BD 4155 Non-vesicular, phenocryst-poor, highly mobile overflow from lava channel from Centre 11, extruded 14.00 h, 22 November 1988.
2.   BD 4167 Phenocryst-poor driblets extruded from end of highly mobile flow from Centre 5, extruded 11.30 h, 24 November 1988.
3.   BD 4159 Non-vesicular, highly mobile flow from Centre 5, extruded 07.00 h, 24 November 1988.
4.   BD 4157 Non-vesicular, highly mobile flow from central hornito (approximately equidistant from Centre 5 and Centre 11), extruded during night of 22–23 November 1988).
5.   BD 4152 Scoriaceous lava, north flank of Centre 11, extruded ? 20–21 November 1988.
6.   BD 4156 Pahoehoe lava in Southern Depression, extruded 14.00 h, 22 November 1988.
7.   BD 4169 Highly vesicular, viscous lava, frothing lava pool, Hornito 5, collected 15.00 h, 24 November 1988.
8.   BD 4151 Scoriaceous lava, collected 24–25 June 1988. Donated by C. Nyamweru.
A,B BD 114 Pahoehoe lava and BD 118 Aa lava, extruded October 1960 (calculated H_2O-free; Dawson 1962 a).

Na/K, K/Rb and Ca/Sr ratios and higher Rb/Sr ratios in the aphyric matrix-rich lava. The same chemical features pertain to aphyric lava BD4155 from centre 11; from this we infer that K_2O, BaO, the halogens and Rb are concentrated preferentially within the matrix. The same applies to the rare earth elements (below).

Table 8. Trace elements concentrations and element ratios in 1988 natrocarbonatite lavas

Sample	1	2	3	4	5	6	7	8
V	232	251	164	178	190	163	134	184
Cu	11	11	10	14	9	9	7	14
Zn	303	43	61	119	83	160	136	93
Rb	259	284	212	237	243	221	193	225
Y	15	<5	5	7	<5	10	<5	9
Zr	<20	20	28	29	<20	30	<20	27
Nb	74	13	15	17	16	46	14	19
Pb	212	47	69	112	79	132	108	84
K/Rb	270	251	303	284	280	292	312	357
Rb/Sr × 10^3	17.6	18.8	14.9	16.5	16.8	15.4	14.0	12.7
Ca/Sr	6.50	6.18	7.12	6.88	6.71	7.08	7.59	6.86
Na_2O/K_2O	3.68	3.77	4.14	3.96	3.88	4.17	4.50	4.51

Table 9. Rare earth element and yttrium concentrations of the carbonatite lavas

	1	2	3	4	5	6	7	8
La	789	669	580	572	570	640	497	615
Ce	860	672	616	603	582	700	543	669
Pr	62	57	49	46	42	52	42	54
Nd	142	107	101	95	90	118	90	112
Sm	11.1	5.8	6.3	6.2	6.2	8.7	5.8	7.5
Eu	2.30	1.14	1.22	1.15	1.13	1.81	1.10	1.48
Gd	6.6	6.4	4.3	3.8	3.7	5.3	3.7	5.2
Dy	3.35	1.80	1.80	1.65	1.70	2.55	1.55	2.20
Ho	0.60	0.45	0.4	0.35	0.34	0.48	0.34	0.48
Yb	0.70	0.35	0.35	0.30	0.35	0.52	0.30	0.45
Lu	0.08	0.09	0.08	0.04	0.05	0.06	0.06	0.07
Y	15.6	6.7	7.0	6.1	6.5	11.1	5.9	9.1

Specimen numbers as in Table 7.
Analyses by I.C.P. (NERC/I.C.P. Unit, Royal Holloway and Bedford New College).

Whereas most of the range in compositions can be attributed to differing phenocryst/matrix ratios, there are other small differences in the compositions of lavas from Centres 5 and 11. Those from Centre 11 contain more total iron, Mn, Zn, Nb and Pb; there are positive correlations between Fe and these other elements in the Centre 11 lavas, suggesting that the difference may be due to a higher content of the sulphide and spinel in the Centre 11 lavas, a feature not readily appreciated during petrographic examination due to their fine grain size. The highest concentrations are in BD4155, that quenched close to the parent magma pool; it contains up to four times the concentrations in BD4156 and BD4152 that consolidated more distally. The lowest concentrations are found in driblets at the distal toe of a Centre 5 flow (BD4167). As these elements are

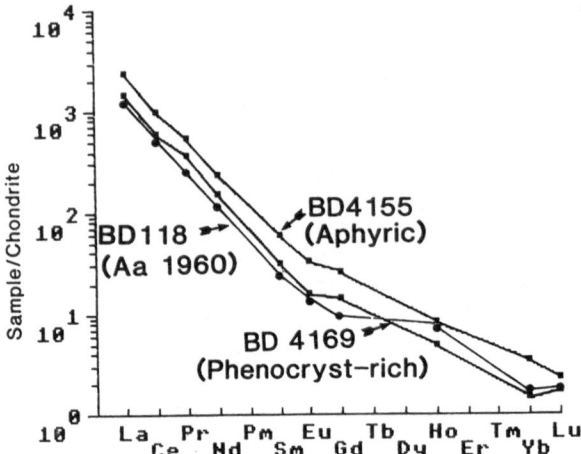

Fig. 3. Chondrite-normalized plots for aphyric and phenocryst-rich lavas from the November 1988 eruption, and a phenocryst-rich lava (BD118) erupted in 1960. Normalized against values given by Wakita et al. (1971). Plots for other November 1988 lavas lie between those for BD4155 and BD4169

mainly in the sulphide, the relatively low concentrations in the distal samples may be due to settling out of the dense sulphide grains during flow.

Rare-earth element analyses (Table 9) also show a range in concentrations, with highest values (ca. 1800 ppm) in aphyric lavas, and the lowest (ca. 1200 ppm) in phenocryst-rich BD4169; concentrations of REE in the other samples fall between these two extremes. Despite differences in absolute concentrations, chondrite-normalized plots for the two extreme lavas (Fig. 3) show very similar patterns, with a very steep slope between La and Eu, and a shallower slope between Gd and Lu; there are no anomalies. Plots for other samples plot between those for BD4155 and 4169. Light REE concentrations are very high (between 1×10^3 and $2 \times 10^3 \times$ chondrite values), whilst those for the heavy REE are $<10 \times$ chondrite. In both their concentrations and plot shape, the more porphyritic lavas are similar to a phenocryst-rich aa lava extruded in October 1960 (Fig. 3).

4.1 Isotope Geochemistry

4.1.1 Stable Isotopes

Analyses of the isotopes of carbon, oxygen and sulphur are given in Table 10. The values for carbon and oxygen, together with those for 1960 lavas (O'Neil and Hay 1973; Sheppard and Dawson 1973), two 1985 lavas (Hay 1989) and a June 1988 sample (Javoy et al. 1988), fall within the limits established for primary, unaltered carbonatites (Sheppard and Dawson 1973). These isotopes are discussed more fully by Keller and Hoefs (Chap. 8, this Vol.).

Table 10. Stable isotope analyses

Sample	$\delta^{13}C$ (PDB) (permil)	$\delta^{18}O$ (SMOW) (permil)
BD 4151	−6.7	+6.2
BD 4159	−6.6	+6.0
BD 4166	−6.1	+5.9

1σ error ± 0.02 per mill. Analyst D. Jackson.

		$\delta^{34}S$ (CDT) (permil)
BD 4159	Sulphide	+2.8
BD 4159	Sulphate	+9.0

1σ error ± 0.2 per mill. Analyst: A.E. Fallick.

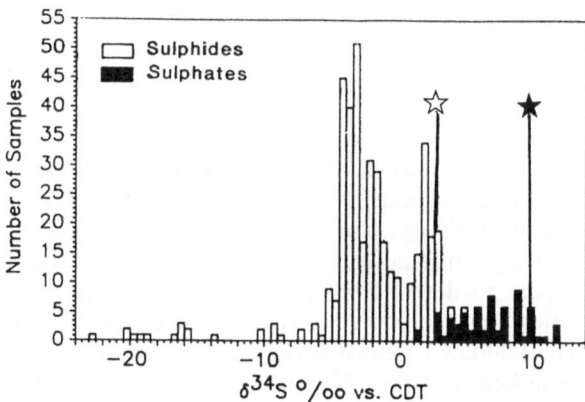

Fig. 4. Sulphur isotope ratios for sulphate and sulphide in lava BD4155 (*stars*), compared with other carbonatite data compiled by Deines (1989)

Sulphur isotope analyses (Table 10) have been made on specimen BD4159, on both a sulphide concentrate and on the acid-soluble fraction in which the sulphur derives from the sulphate molecule within the nyerereite and gregoryite. Both contain heavier sulphur than the meteoritic value (Fig. 4), and plot towards the heavy end of the established $d^{34}S$ ranges for suphides and sulphates from other carbonatites (reviewed by Deines 1989). The sulphide value of 2.8‰ is within the range of 1 to 5‰ found in high-temperature carbonatites such as Phalaborwa and Eastern Sayan, and which is believed to represent mantle values (Mitchell and Krouse 1975). However, the Oldoinyo Lengai sulphide differs in co-existing with phases containing oxidized sulphur, in contrast to the high-temperature plutonic carbonatites that contain no sulphate. These high values could be due to either vapour loss of light ^{32}S, or contamination of the magma by heavy crustal sulphur. Because the carbon and isotope ratios show no evidence of crustal contamination, we incline to the former alternative; certainly, sulphur transport in the gas phase is shown by the presence of 0.15% H_2S in the gases co-existing with the lava (Javoy et al. 1988), and deposition of sulphur around the vents.

4.2 Radiogenic Isotopes

Two of the objectives of radiogenic isotope studies on lavas erupted in rift zones are to ascertain the type of mantle in which the magma originates and to search for evidence of interaction with crustal materials during magma ascent. In the case of the Oldoinyo Lengai carbonatites, there are potential interpretational problems because of the intrisically high amounts of Sr, Pb and Nd ($>1.5\%$, 90–120 ppm, 47–212 ppm respectively) which can make the rocks insensitive to even moderate degrees of contamination by material such as the average lower crust, for which estimates of concentrations are: Sr 230 ppm, Pb 4 ppm, Nd 13 ppm (Taylor and McLennan 1985)

4.2.1 Nd and Sr Isotopes

Previously published Nd and Sr isotope data for Oldoinyo Lengai carbonatites (Bell and Blenkinsop 1987; Keller and Krafft 1990) fall close to the regression line through age-corrected data for Ugandan carbonatites, termed the East African Carbonatite Line (Bell and Blenkinsop 1987). The natrocarbonatites appear to derive from a region with higher time-averaged Rb/Sr and Nd/Sm ratios than the depleted MORB source; in this respect they are similar to OIB. The Oldoinyo Lengai rocks show no evidence of the enhanced ^{87}Sr/^{86}Sr ratios exhibited by young Naivasha basalts, which has been attributed to interaction with lower crustal amphibolite (Davies and Macdonald 1987). For a given ^{87}Sr/^{86}Sr ratio, the Oldoinyo Lengai carbonatites contain less radiogenic Nd than other East African carbonatites, and show no evidence for interaction with lower crustal granulites (Cohen et al. 1984).

4.2.2 Pb Isotopes

Previous work on African carbonatites (Lancelot and Allègre 1974; Grünenfelder et al. 1986) reported Pb isotope compositions similar to those of Atlantic alkali basalts, with the carbonatites being interpreted as mantle-derived, but with some contamination with crustal radiogenic Pb. The lead in North American carbonatites (Kwon et al. 1989) also has isotopic compositions consistent with generation from OIB-type mantle, but with little evidence for interaction with the sub-continental lithosphere. New Pb isotope data for three Oldoinyo Lengai carbonatites from the November 1988 eruption are given in Table 11, and are

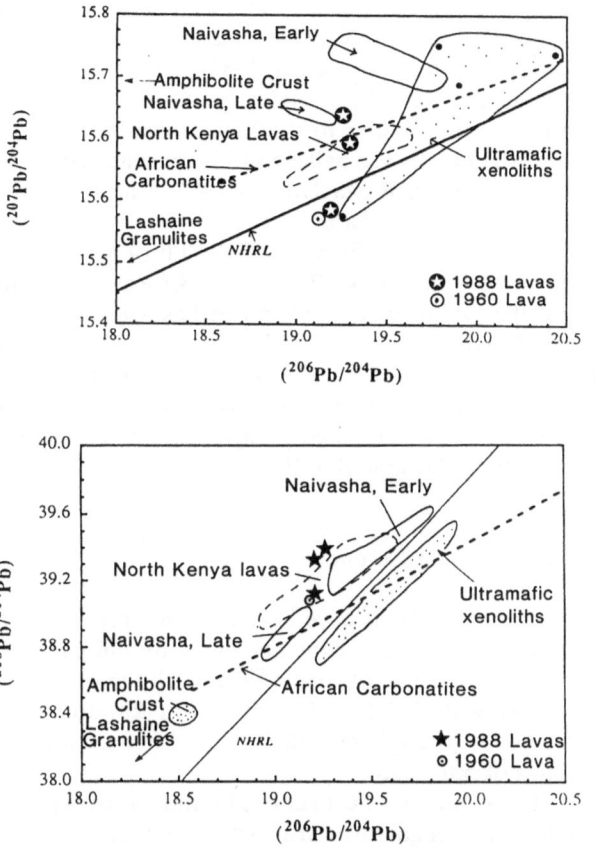

Fig. 5. a $^{207}Pb/^{204}Pb$ vs. $^{206}Pb/^{204}Pb$ isotope plot for Oldoinyo Lengai carbonatites. The *1960* carbonatite analysis is from Williams et al. (1986). *NHRL* is the Northern Hemisphere Regression Line that represents present-day Pb isotope composition of the mantle (Hart 1984). The *African Carbonatites* line is based on data from Lancelot and Allègre (1974) and Grünenfelder et al. (1986). Other data sources are: Tanzanian *ultramafic xenoliths* and *Lashaine granulites* Cohen et al. (1984); *North Kenya basalts* Norry et al. (1980); *Naivasha basalts* Davies and Macdonald (1987). **b** $^{208}Pb/^{204}Pb$ vs. $^{206}Pb/^{204}Pb$ istope plot for Oldoinyo Lengai carbonatites. Data sources as in **a**

plotted on combined Pb isotope diagrams (Fig. 5a,b); also plotted are data from Williams et al. (1986) for a 1960 lava that has a composition very close to BD4159. The available Pb-isotope data for Pleistocene to Recent basalts and trachytes from northern and central Kenya (Norry et al. 1980; Davies and MacDonald 1987) are shown for comparison, together with data from mafic granulites and ultramafic xenoliths from Lashaine and Pello Hill, two Tanzanian Recent tuff cones (Cohen et al. 1984); the ultramafic xenoliths are samples of the subcontinental lithosphere beneath N. Tanzania.

In Fig. 5a, the Oldoinyo Lengai carbonatite data fall on a steep array that cuts across both the Northern Hemisphere Regression Line (NHRL – Hart

1984) and the African carbonatite regression line (Lancelot and Allègre 1974). The NHRL is effectively the mean isotopic composition of Atlantic MORB and North Atlantic OIB (Hart 1984). The Oldoinyo Lengai samples are not as radiogenic as the Early Naivasha basalts, but have $^{207}Pb/^{204}Pb$ and $^{206}Pb/^{204}Pb$ ratios broadly similar to the Late Naivasha and north Kenya basalts. The Early Naivasha basalts are believed to be partially contaminated by amphibolitic lower crust (Davies and MacDonald 1987), the (calculated) composition of which is also shown in Fig. 5a,b.

On the $^{208}Pb/^{204}Pb$ vs. $^{206}Pb/^{204}Pb$ plot (Fig. 5b), the Oldoinyo Lengai data fall above both the NHRL and the African carbonatite regression line, indicating that the Oldoinyo Lengai source has a higher time-integrated Th/U ratio than the MORB source. The Oldoinyo Lengai data are, as in Fig. 5a, broadly similar to the north Kenya basalts and to some of the Naivasha basalts (both Early and Late). There is no evidence from Fig. 5a for interaction with lower crustal material as the Oldoinyo Lengai data do not trend in the direction of either the amphibolite or granulite fields.

4.2.3 Thorium Isotopes

Several of the lavas have been analyzed for U and Th decay series nuclides (Pyle et al. 1991). In particular, the lavas are unique in showing disequilibria between ^{228}Th-^{232}Th and between ^{228}Ra-^{232}Th, and are enriched in (^{238}U) relative to (^{230}Th) with U/Th weight ratios of 2.0 to 3.2. Driblets, the most extreme fractionates, are highly enriched in U and Ra relative to Th, with U/Th weight ratios of 5.6 to 6.4.

The combined Pb (Table 11) and Th isotope data (Pyle et al. 1991) for the Oldoinyo Lengai lavas are compared with data for MORB, OIB and continental volcanics (Allègre et al. 1986, Gill et al. 1992) in Fig. 6. K_{Th} represents the instantaneous Th/U weight ratio of the source region today. The ratio $^{208}Pb^*/^{206}Pb^*$, where $^{208}Pb^*$ and $^{206}Pb^*$ are the amounts of radiogenic lead (normalized to ^{204}Pb) produced since the Earth formed (see caption for Fig. 6), is a measure of the time-integrated Th/U weight ratio of a sample, alternatively referred to as K_{Pb} (Oversby and Gast 1968). The locus of points where $K_{Th} = K_{Pb}$ is the geochron. Most young basalts are derived from source regions with Th/U ratios

Table 11. Pb isotope data from Oldoinyo Lengai

Sample	$^{208}Pb/^{204}Pb$	$^{207}Pb/^{204}Pb$	$^{206}Pb/^{204}Pb$
BD 4151	39.365	15.676	19.245
BD 4157	39.396	15.637	19.261
BD 4159	39.125	15.552	19.200
2σ error[a]	±0.058	±0.024	±0.029

[a] Including counting statistics, and reproducibility. All ratios corrected for fractionation. Analyst P. Browning.

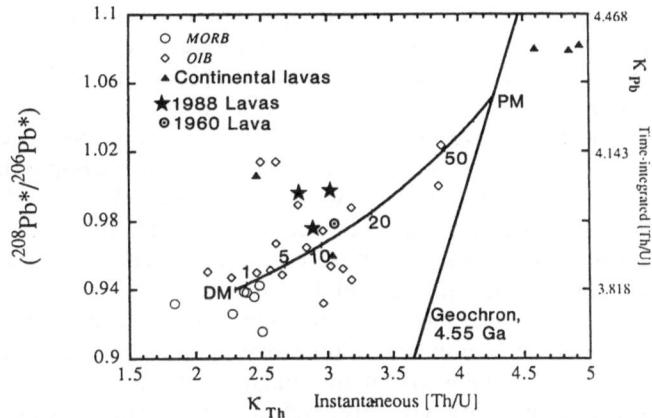

Fig. 6. Combined-Pb isotope and Th-isotope plot for Oldoinyo Lengai carbonatites, MORB, OIB and continental volcanics. K_{Th} is defined as 3.056. $(^{232}Th/^{230}Th)$, where $(^{232}Th/^{230}Th)$ is an activity ratio (Capaldi et al. 1982). The y axis K_{PB} is defined as 4.062. $(^{208}Pb*/^{206}Pb*)$. $^{208}Pb* = (^{208}Pb/^{206}Pb)$-29.476 and $^{206}Pb* = (^{206}Pb/^{204}Pb)$-9.307 (Tatsumoto et al. 1973; Oversby and Gast 1968). Data from Oversby and Gast (1968), Newman et al. (1983, 1984), Allègre et al. (1986), Williams et al. (1986), Gill et al. (1992) and Pyle et al. (1991). *MORB* and *OIB* White and Hoffman (1982), Zindler et al. (1982)

lower than their time-integrated ratio. The plot shows that the Oldoinyo Lengai carbonatites were generated in an upper mantle region with Th-U-Pb isotopic characteristics similar to OIB. This region must have been depleted in Th relative to U within the last 1 billion years, but does not have a Th/U ratio as low as that of the MORB source. Also shown is a curve for the mixing of depleted mantle (DM) and primitive (unfractionated) mantle (PM) with compositions as defined by Zindler and Hart (1986); a possible interpretation would be that the source region for the U, Th and Pb in the Oldoinyo Lengai carbonatites is a mixture of 90% depleted mantle with 10% of an unfractionated component.

5 Discussion

The alkali carbonate lavas consist of phenocrysts of nyerereite and gregoryite in a fine-grained, rapidly chilled, groundmass. Since the first analyses of the phenocrysts (McKie and Frankis 1977; Gittins and McKie 1980), it has been recognized that there is widespread substitution of K for Na; Sr and Ba for Ca; and S, P, F and Cl for C in the lattices of these complex carbonates. They are essentially solid solutions in the system Na_2CO_3-K_2CO_3-$CaCO_3$; although the perthitic texture and lamellae of nyerereite in gregoryite in some specimens indicate the incipient breakdown of the gregoryite solid solution. The minerals of the groundmass, in which most of the Fe, Mn, Mg, REE and halogens, and considerable K_2O, BaO and SrO are concentrated, also show certain solid solutions that are very rare or have not been reported previously. Cases in point are the Mn, Fe sulphides, the Fe, Mn spinel, the Na-bearing sylvite and the Ca-Ba-Sr-Na-K ?carbonate (phase X). The high temperature of formation is presumed to have given rise to these solid solutions, because crystallization of these chemistries at low temperatures, for example in a hydrothermal or sedimentary environment, would have given rise to end-members of the solid solutions. Besides the presence of the inherently alkali-rich phases, such as sylvite, the high-alkali nature of the groundmass is reflected in the high alkali content of the apatite and the spurrite. The presence of spurrite (a mixed carbonate-silicate) reflects the high fCO_2, the carbonate melt mimicking the thermally metamorphosed limestone in which spurrite is normally found.

The bulk chemistry of the magma indicates that it is an ionic melt. This is reflected in its low viscosity, and may be presumed also to give rise to a greatly enhanced ability of the ions to migrate within the melt, particularly compared with polymerized silicate melts. The melt structure, combined with halogen concentration, might enable ion mobility to extend to the point when very localized residual melt patches crystallize, giving rising to such features as the sylvite-fluorite-"nyerereite" intergrowths and the widely variable compositions of the patches of phase X. Enhanced ion mobility in the sub-solidus may also be the reason why, despite rapid chilling, the lavas contain no glass.

The composition and style of extrusion of the 1988 lavas show a reversion to the quiet extrusions of dominantly carbonate lavas, such as took place in the early 1960s prior to the violent eruption of carbonate-silcate ashes in 1966. Dawson et al. (1992) suggest that the eruption of the mixed ashes in August 1966, following a period of prolonged extrusion of carbonate magma, might indicate the mixing of silicate and carbonate fractions of a density-stratified magma chamber of which the carbonate lavas extruded in 1960–1966 represent the low-density (? immiscibly separated) upper part.

For the 1988 lavas, the disequilibria between (^{232}Th)-(^{228}Ra)-(^{228}Th) (Pyle et al. 1991) are consistent with two models for timing of the magma genesis: (1) if the magma extruded in 1988 was formed at one instant, then that was 20 ± 1 years before April 1989 (the date of measurement), i.e. 1969 ± 1 year; or (2) if there is a recharged, steady-state magma chamber beneath the volcano, then the magma residence time is 81 ± 9 years. Model 1 is consistent with new magma formation shortly after the volcano was eviscerated by the 1966 and 1967 eruptions. Model 2 does not relate to any verified eruption.

Carbonatites, being enriched in Sr, Nd and Pb, are robust carriers of the isotopic signature of their ultimate source region. There is no evidence from the available data that the Oldoinyo Lengai carbonatites have taken any of the isotopic characteristics of the crust through which they have passed. Their Pb, Nd and Sr isotopic compositions suggest that the carbonatites or their parental melts were formed by melting of a portion of the mantle or subcontinental lithosphere. The isotopic composition of the Oldoinyo Lengai carbonatites converges closely with the few data available for young mafic lavas from the Gregory Rift Valley; clearly, many more data are needed. Compared with MORB, the carbonatites have more radiogenic $^{207}Pb/^{204}Pb$ and $^{87}Sr/^{86}Sr$ ratios.

In many respects they are most similar to OIB and fall within the bounds of the southern hemispheric "DUPAL" anomaly (Dupré and Allègre 1983; Hart 1984).

Acknowledgements. The expedition to Oldoinyo Lengai was sponsored by the Royal Society of London, and Gill Norton's participation was made possible by a special grant from the Natural Environment Research Council. In East Africa, the logistics of the expedition were expertly handled by the Peterson brothers of Dorobo Safaris and their staff. We are especially grateful to Professor Celia Nyamweru for arranging many things that smoothed the path of the expedition and for her many kindnesses and personal hospitality. Stuart Kearns, University of Edinburgh, assisted with the microprobe analyses, and Fergus Gibb, University of Sheffield, helped with the analyses of nyerereite. Alan Saxby, University of Sheffield, made the bulk rock analyses. The REE analyses were made at the NERC-funded ICP unit at Royal Holloway and Bedford New College; the Isotope Geology Unit at SURRC is supported by NERC and the Scottish Universities. The stable isotope analyses were made in the laboratory of Dr. C. Pillinger. D.M. Pyle thanks St Catherine's College for a Research Fellowship, Caltech for a Visiting Associateship, and Ed Stolper for use of his computing facilities.

References

Allègre CJ, Dupré B, Lewin E (1986) Thorium/uranium ratio of the earth. Chem Geol 56:219–227

Bell K, Blenkinsop J (1987) Nd and Sr isotopic compositions of East African carbonatites: implications for mantle heterogeneity. Geology 15:99–102

Capaldi G, Cortini M, Pece R (1982) Th isotopes at Vesuvius: evidence for open-system behaviour of magma-forming processes. J Volcanol Geotherm Res 14:247–260

Clarke DB (1979) Synthesis of nickeloan djerfisherites and the origin of potassic sulphides at the Frank Smith Mine. In: Boyd FR, Meyer HOA (eds) The mantle sample: inclusions in kimberlites and other volcanics. Am Geophys Union, Washington DC, pp 300–308

Cohen RS, O'Nions RK (1982) The lead, neodymium and strontium isotopic structure of ocean ridge basalts. J Petrol 23:299–324

Cohen RS, O'Nions RK, Dawson JB (1984) Isotope geochemistry of xenoliths from East Africa: implications for development of mantle reservoirs and their interactions. Earth Planet Sci Lett 68:209–220

Colville AA, Colville PA (1977) Paraspurrite, a new polymorph of spurrite from Inyo County, California. Am Mineral 62:1003–1005

Czamanske GK, Lanphere MA, Erd RC, Blake MC (1978) Age measurements of potassium-bearing sulfide minerals by the ^{40}Ar/^{39}Ar technique. Earch Planet Sci Lett 40:107–110

Czamanske GK, Erd RC, Sokolova MJ, Dobrovolskaya MJ, Dmitrieva MT (1979) New data on rasvumite and djerfisherite. Am Mineral 64:776–778

Davies GR, Macdonald R (1987) Crustal influences in the petrogenesis of the Naivaska basalt-comendite complex: combined trace element and Sr-Nd-Pb isotope constraints. J Petrol 28:1009–1031

Dawson JB (1962a) Sodium carbonate lavas from Oldoinyo Lengai, Tanganyika. Nature 195:1075–1076

Dawson JB (1962b) The geology of Oldoinyo Lengai. Bull Volcanol 24:349–387

Dawson JB, Bowden P, Clark GC (1968) Activity of the carbonatite volcano Oldonyo Lengai, 1966. Geol Rundsch 57:865–879

Dawson JB, Pinkerton H, Norton GE, Pyle DM (1990) Physicochemical properties of alkali carbonatite lavas: data from the 1988 eruption of Oldoinyo Lengai, Tanzania. Geology 18:260–263

Dawson JB, Smith JV, Steele IM (1992) 1966 ash eruption of Oldoinyo Lengai: mineralogy of lapilli, and mixing of carbonatite and silicate magma. Min Mag 56:1–16

Deines P (1989) Stable isotope variations in carbonatites. In: Bell K (ed) Carbonatites – genesis and evolution. Unwin Hyman. London, pp 301–359

Du Bois CGB, Furst, Guest NJ, Jennings DJ (1963) Fresh natrocarbonatite lava from Oldoinyo L'Engai. Nature 197:445–446

Dupré B, Allègre CJ (1983) Pb-Sr isotope variation in Indian Ocean basalts and mixing phenomena. Nature 303:142–146

Gill JB, Pyle DM, Williams RW (1992) Igneous rocks. In: Ivanovich M, Harmon RS (eds) Uranium-series disequilibrium: applications to earth, marine and environmental sciences. Oxford University Press, Oxford, pp 207–258

Gittins J, McKie D (1980) Alkalic carbonatite magmas: Oldoinyo Lengai and its wider applicability. Lithos 13:213–215

Grünenfelder MH, Tilton GR, Bell K, Blenkinsop J (1986) Lead and strontium isotope relationships in the Oka carbonatite complex, Quebec. Geochim Cosmochim Acta 50: 461–468

Hart SR (1984) A large-scale isotope anomaly in the Southern Hemisphere mantle. Nature 309:753–757

Hay RL (1989) Holocene carbonatite-nephelinite tephra deposits of Oldoinyo Lengai, Tanzania. J Volcanol Geotherm Res 37:77–91

Hogarth DD (1989) Pyrochlore, apatite and amphibole: distinctive minerals in carbonatite. In: Bell K (ed) Carbonatites – genesis and evolution. Unwin Hyman, London, pp 105–148

Javoy M, Pineau F, Cheminée JL, Krafft M (1988) The gas magma relationship in the 1988 eruption of Oldoinyo Lengai (Tanzania). EOS 69:1466

Keller J, Krafft M (1989) Composition of natrocarbonatite lavas, Oldoinyo Lengai 1988. TERRA Abstr 1:286

Keller J, Krafft M (1990) Effusive natrocarbonatite activity of Oldoinyo Lengai, June 1988. Bull Volcanol 52:629–645

Kjarsgaard BA, Hamilton DL (1989) The genesis of carbonatites by immiscibility. In: Bell K (ed) Carbonatites – genesis and evolution. Unwin Hyman, London, pp 388–404

Krafft M, Keller J (1989) Temperature measurements in carbonatite lava lakes and flows from Oldoinyo Lengai, Tanzania. Science 245:168–170

Kwon ST, Tilton GR, Grünenfelder MH (1989) Lead isotope relationships in carbonatites and alkalic complexes: an overview. In: Bell K (ed) Carbonatites – genesis and evolution. Unwin Hyman, London, pp 360–387

Lancelot JR, Allègre CJ (1974) Origin of carbonatitic magma in the light of the Pb-U-Th isotope system. Earth Planet Sci Lett 22:223–238

Mason B (1947) Mineralogical aspects of the system Fe_3O_4-Mn_3O_4-$ZnMn_2O_4$-$ZnFe_2O_4$. Amer Mineral 32:426–447

McKie D, Frankis EJ (1977) Nyerereite: a new volcanic carbonate mineral from Oldoinyo L'engai. Tanzania. Z Kristallogr 145:73–95

Mitchell RH, Krouse HR (1975) Sulphur isotope geochemistry of carbonatites. Geochim Comochim Acta 39:1505–1513

Newman S, Finkel RC, MacDougall JD (1983) ^{230}Th-^{238}U disequilibrium systematics in oceanic tholeiites from 21°N on the East Pacific Rise. Earth Planet Sci Lett 65:17–33

Newman S, Finkel RC, MacDougall JD (1984) Comparison of ^{230}Th-^{238}U disequilibrium systematics in lavas from three hotspot regions Hawaii, Prince Edward and Samoa. Geochim Cosmochim Acta 48:315–324

Norry MJ, TrucklePH, Lippard SJ, Hawkesworth CJ, Weaver SD, Marriner GF (1980) Isotopic and trace element evidence from lavas, bearing on mantle heterogeneity beneath Kenya. Philos Trans R Soc Lond A297:259–271

Nyamweru C (1988) Activity of Ol Doinyo Lengai volcano, Tanzania, 1983–1987. J Afr Earth Sci 7:603–610

O'Neil JR, Hay RL (1973) $^{18}O/^{16}O$ ratios in cherts associated with the saline lake deposits of East Africa. Earth Planet Sci Lett 19:257–266

Oversby V, Gast PW (1968) Lead isotope composition and uranium decay series disequilibrium in recent volcanic rocks. Earth Planet Sci Lett 5:199–206

Peterson TD (1990) Petrology and genesis of natrocarbonatite. Contrib Mineral Petrol 105: 143–155

Pyle DM, Dawson JB, Ivanovich M (1991) Short-lived decay series disequilibria in the natro-carbonatite lavas of Oldoinyo Lengai, Tanzania: constraints on the timing of magma genesis. Earth Planet Sci Lett 105:378–396

Ramdohr P (1957) Eisenalabandin, ein merkwurdiger natürlicher Hochtemperatur-Mischkristall. N Jahrb Mineral Abh 91:89–93

Rønsbo JG (1989) Coupled substitutions involving REEs and Na and Si in apatites in alkaline rocks from the Ilimaussaq intrusion, South Greenland, and the petrological implications. Am Mineral 74:896–901

Sheppard SMF, Dawson JB (1973) $^{13}C/^{12}C$ and D/H isotope variations in primary igneous carbonatites. Fortsch Mineral 50:128–129

Shive PN, Nyblade AA, Wittke JH (1990) Magnetic properties of some carbonatites from Tanzania, East Africa. Geophys J Int 103:103–109

Skinner BJ, Luce FD (1971) Solid solutions of the type (Ca, Mg, Mn, Fe)S and their use as geothermometers for the enstatite chondrites. Am Mineral 56:1269–1296

Tatsumoto M, Knight RJ, Allègre CJ (1973) Time difference in the formation of meteorites as determined from the ratio of lead-207 to lead-208. Science 180:1278–1283

Taylor SR, McLennan SM (1985) The continental crust: its composition and evolution. Blackwell, Oxford, 312 pp

Uhlig C (1907) Der sogennante grosse Ostafrikanische Graben zwischen Magad (Natron See) und Lawa ya Mueri (Manyara See). Geogr Zeit 15:478–505

Wakita H, Rey P, Schmitt RA (1971) Abundances of the 14 rare-earth elements and 12 other trace elements in Apollo 12 samples, five igneous and one breccia rock and four soils. Proc 2nd Lunar Sci Conf, Houston, Texas, pp 1319–1329

White WM, Hofmann AW (1982) Sr and Nd isotope geochemistry of oceanic basalts and mantle evolution. Nature 296:821–825

Williams RW, Gill JB, Bruland KW (1986) Ra-Th disequilibria systematics: timescale of carbonatite magma formation at Oldoinyo Lengai volcano, Tanzania. Geochim Cosmochim Acta 50:1249–1259

Zindler A, Hart SR (1986) Chemical geodynamics. Ann Rev Earth Planet Sci 14:493–571

Zindler A, Jagoutz E, Goldstein S (1982) Nd, Sr and Pb isotope systematics in a three component mantle: a new perspective. Nature 298:519–523

The Trace Element Composition and Petrogenesis of Natrocarbonatites

J. KELLER[1] and B. SPETTEL[2]

Abstract

Trace element data, including the REE, are reported for natrocarbonatites collected during the 1988 effusive period of Oldoinyo Lengai, Tanzania. These data, together with trace element data for the phenocryst phases nyerereite and gregoryite and for a related peralkaline nephelinite, provide the basis for discussing petrogenetic hypotheses related to alkalic carbonatites. Natrocarbonatites are characterized, compared to the more common calciocarbonatites, dolomitic carbonatites, and ferrocarbonatites, by extreme fractionation of the REE, with $(La/Sm)_N$ ratios >40, high concentrations for Sr, Ba, Rb, K, unusually high contents of Mo, W, U, As and Sb and strong depletion in Ta, Nb, Hf, Zr, Ti, Y, Yb and Lu. The trace element signature of natrocarbonatite as compared with the more common alkali-poor carbonatites shows that Lengai lavas represent exotic compositions so far restricted to this volcano, and cannot be used to model petrogenetic schemes for carbonatites in general. Liquid immiscibility is a viable model for the derivation of natrocarbonatite from an exceptionally peralkaline nephelinite.

1 Introduction

Natrocarbonatites from the recent eruptions of Oldoinyo Lengai are unique in their alkali contents compared to other carbonatites of the East African Rifts and to carbonatites in general. Most carbonatites are calcitic calciocarbonatites (Woolley and Kempe 1989), and less commonly dolomitic (magnesiocarbonatites) or Fe-rich (ferrocarbonatites).

Oldoinyo Lengai erupted natrocarbonatite lavas in an effusive period commencing in 1960 (Dawson 1962). Explosive eruptions in 1966 and 1967 terminated this type of effusive activity. Due to the deliquescent and hygroscopic nature of the natrocarbonatite lavas, no fresh samples were available for geochemical studies until effusive activity resumed and sampling became possible during 1988 (Keller and Krafft 1989, 1990; Dawson et al. 1990).

Because Oldoinyo Lengai is the only example of an active volcano erupting carbonatitic lavas, a major question involves the extent to which the composition and evolution of natrocarbonatites relate to carbonatite petrogenesis in

[1] Institut für Mineralogie, Petrologie und Geochemie, Universität Freiburg, Albertstr. 23b, 79104 Freiburg, Germany
[2] Max-Planck-Institut für Chemie, Abt. Kosmochemie, Saarstr. 23, 55020 Mainz, Germany

general (Le Bas 1987; Gittins 1989; Keller 1989; Kjarsgaard and Hamilton 1989; Peterson 1990). It is widely agreed that carbonatites, and also the natro-carbonatites of Oldoinyo Lengai, are derived ultimately from the mantle (Bell and Blenkinsop 1989; Wyllie et al. 1990). The primary igneous C and O isotopic ratios, in particular, support this inference (Javoy et al. 1989; Keller and Hoefs, Chap. 8, this Vol.), as well as the isotopic compositions of Sr and Nd (Bell and Dawson, Chap. 7, this Vol.).

Different mechanisms for the separation of the carbonatite melt from its silicate parent have been proposed. A liquid immiscibility relationship between natrocarbonatites and peralkaline nephelinites (Peterson 1989a,b, 1990; Keller and Krafft 1990) is supported by earlier experimental results of Koster van Groos and Wyllie (1968) and Freestone and Hamilton (1980). Wyllie et al. (1990) concluded that liquid immiscibility is unlikely to occur at mantle pres-sures, but may occur in carbonated silicate melts at lower pressures producing conjugate immiscible silicate-carbonatitic liquids. Kjarsgaard et al. (Chap. 12, this Vol.) and Peterson and Kjarsgaard (Chap. 11, this Vol.) have presented new experimental evidence and phase relationships supporting the low-pressure immiscible separation of natrocarbonatite from highly peralkaline nephelinites. Twyman and Gittins (1987) and Gittins (1989) have challenged the liquid immiscibility hypothesis and consider natrocarbonatites as a late-stage fractiona-tion product of less alkaline calciocarbonatites. Wallace and Green (1988) have shown experimentally that a primary carbonatite partial melt in equilibrium with pargasite lherzolite is a Na-rich (5 wt%) dolomitic carbonatite which could further evolve towards a composition similar to natrocarbonatite of Oldoinyo Lengai (see also Sweeney et al., Chap. 13, this Vol.).

2 Composition and Petrogenesis of Natrocarbonatites

Natrocarbonatites are, in general, highly porphyritic lavas with phenocrysts of the two alkali carbonates, nyerereite and gregoryite (Gittins and McKie 1980; Dawson et al. 1989, 1990; Keller and Krafft 1989, 1990; Peterson 1990). Rapid matrix crystallization has led to a mineralogically complex groundmass com-posed of nyerereite, gregoryite, fluorite and sylvite, with minor Fe-alabandite, apatite, rare sellaite (MgF_2), witherite ($BaCO_3$) and rarer phases, mainly oxides, halides and carbonates (Keller and Krafft 1990; Dawson et al., Chap. 4, this Vol.; Koberski and Keller, Chap. 6, this Vol.). Two petrographic varieties of natrocarbonatite, a crystal-rich porphyritic "normal" variety and a phenocryst-poor to aphyric facies have been described by Keller and Krafft (1990). The aphyric variety represents a residual liquid separated from its phenocryst mush by filter pressing during the effusive processes, and is the most fractionated natrocarbonatite composition of the current activity.

Although the trace element discussion of this study is based on one sample only of each variety, it should be emphasized that the samples for neutron activation analysis were selected out of a larger number of samples for which a very narrow compositional variation had been demonstrated (Keller and Krafft

Table 1. Chemical composition of Oldoinyo Lengai natrocarbonatite lavas, June 1988, of their phenocryst phases nyerereite and gregoryite, and of a closely associated peralkaline nephelinite

No.	OL 102	OL 105	NY	GRE	OL 7
SiO_2	0.16	0.21	–	–	43.50
TiO_2	0.02	0.02	–	–	0.95
Al_2O_3	bd	bd	–	–	13.89
Fe_2O_3	0.28	0.48	0.03	0.06	5.45
FeO	–	–	–	–	2.96
MnO	0.38	0.60	0.10	0.12	0.33
MgO	0.38	0.52	–	–	0.90
CaO	14.02	12.86	23.84	7.48	10.97
SrO	1.42	1.43	2.25	0.61	0.28
BaO	1.66	2.17	0.81	0.27	0.18
Na_2O	32.22	30.42	23.56	45.67	12.00
K_2O	8.38	9.14	8.39	3.82	4.87
P_2O_5	0.85	0.75	0.56	1.96	0.65
CO_2	31.55	28.12	38.96	35.18	1.30
Cl	3.40	5.18	0.26	0.51	na
F	2.50	4.10	0.06	0.20	0.38
SO_3	3.72	5.58	1.15	4.70	0.31
H_2O_+	0.56	0.44	–	–	1.10
$-O = F,Cl$	−1.82	−2.90	−0.08	−0.20	−0.10
Total	99.68	99.12	99.89	100.38	99.92

$^{87}Sr/^{86}Sr^a$	0.70437	0.70439			0.70434
$^{143}Nd/^{144}Nd$	0.51262	0.51263			0.51264

[a] Sr and Nd isotopic data by K. Bell (Keller and Krafft 1990).

OL 102   Porphyritic lava, June 25, 1988 (Keller and Krafft 1990).
OL 105   Aphyric lava, June 25, 1988 (Keller and Krafft 1990).
NY, GRE Nyerereite and gregoryite phenocrysts of OL 102
         Mineral formulas:
         $(Na_{0.82}K_{0.19})_2(Ca, _{Sr,Ba})_{0.975}(CO_3)_2$ for nyerereite
         $Na_{1.74}K_{0.1}(Ca,_{Sr,Ba})_{0.16}CO_3$ for gregoryite
         both with $(SO_4)^{2-}$, $(PO_4)^{3-}$, F^- and Cl^- for $(CO_3)^{2-}$.
OL 7   Peralkaline combeite-wollastonite nephelinite, western rim lava.

1990). Therefore the data are considered representative for the present activity. The major element composition of the porphyritic variety (OL 102) and the aphyric composition (OL 105) are listed in Table 1, along with electron microprobe data for nyerereite and gregoryite phenocrysts. Trace element data for lavas and separated phenocrysts are reported in Table 2.

Peralkaline, highly evolved nephelinites are considered to play a key role in the origin of natrocarbonatites by liquid immiscibility (Peterson 1989a,b, 1990; Keller and Krafft 1990; Kjarsgaard et al., Chap. 12, this Vol.). Very few detailed chemical data exist until now for these lavas. Keller and Krafft (1990) have drawn attention to the close association of the actual natrocarbonatite activity and a specific nephelinite, the peralkaline wollastonite- and combeite-bearing nephelinitic rim lava (OL 7). This nephelinite is intercalated within the sequence of historical and subhistorical natrocarbonatite flows. Moreover,

Table 2. Trace element composition and REE (in ppm) of natrocarbonatite, of nyerereite and gregoryite, and of peralkaline nephelinite (samples as in Table 1)

No.	OL 102	OL 105	NY	GRE	OL 7
V	116	174	na	na	215
Sc	0.02	0.1	bd	bd	9
Cr	1.5	bd	bd	bd	10
Ni	bd	bd	bd	bd	bd
Ba	14300	19475	7600	3530	1560
Sr	11975	12050	17980	6880	2325
Rb	178	219	185	76	110
Cs	6.4	9.4	3.7	2.2	2.12
Zr	<10	<10	na	na	800
Nb	28	44	na	na	295
Zn	88	151	<15	18	260
Ga	<20	<35	<12	<20	36.4
As	18.0	26.7	3.75	16.3	4.9
Se	<0.5	<0.4	<0.5	<0.5	<1.0
Br	68.7	114	28.8	29.2	1.7
Mo	125	188	<7.0	24.0	<3
W	48.9	77.5	4.4	<10	<4
Sb	2.2	3.5	0.4	0.5	0.8
Hf	0.13	<0.05	<0.1	<0.1	12.4
Ta	<0.025	<0.015	<0.03	<0.1	4.8
Th	3.77	5.78	0.84	0.96	18.1
U	10.6	17.4	2.2	2.5	9.1
La	545	778	408	175	149
Ce	645	859	515	178	243
Nd	102	129	89	33	77.0
Sm	7.80	10.86	7.11	2.43	13.3
Eu	1.62	1.93	1.31	0.33	3.77
Tb	0.32	0.42	0.24	<0.1	1.47
Dy	1.75	1.74	0.7	<1.4	8.18
Yb	0.46	0.44	0.45	<0.1	3.54
Lu	0.058	0.062	0.032	<0.04	0.48
Y	7	12	na	na	46
Na/K	3.51	2.97	2.28	8.56	2.25
K/La	118.5	94.7	158	217	258
Ba/Sr	1.2	1.6	0.4	0.5	0.7
Ba/La	23.4	20.7	18.6	20.2	11.2
La/Yb	1129	1766	907	>1750	42.1
La_N/Sm_N	42.6	43.7	35.0	43.9	6.75
Th/U	0.36	0.33	0.39	0.38	1.99
Th/Ta	>150	>385	>28	>10	3.77
Nb/Ta	>1120	>2900	–	–	61
Zr/Nb	<0.36	<0.23	–	–	2.7

natrocarbonatites and peralkaline nephelinite OL 7 show identical Sr and Nd isotopic ratios (Keller and Krafft 1990; Bell and Dawson, Chap. 7, this Vol.). Given the generally large variations in isotopic composition of Oldoinyo Lengai lavas (Donaldson et al. 1987; Bell and Dawson, Chap. 7, this Vol.) the isotopic evidence suggests that natrocarbonatite and nephelinite OL 7 evolved from the same magma batch. Therefore, Keller and Krafft (1990) have considered this specific nephelinite as a possible candidate for the composition of silicate melts related by liquid immiscibility to natrocarbonatite. Kjarsgaard et al. (Chap. 12, this Vol.) have experimentally demonstrated liquid immiscibility between wollastonite nephelinite with a peralkalinity index <2, and natrocarbonatite, but were unable to document any immiscible relationships between natrocarbonatite and hyperalkaline combeite nephelinite. However, their experiments with this composition were restricted to only two runs, both of which were open-tube experiments. Under the same conditions, even the less peralkaline wollastonite nephelinite composition failed to show any evidence for liquid immiscibility. The wollastonite-combeite nephelinite, OL 7, has a bulk rock peralkalinity index of 1.8. On the basis of its composition (Tables 1 and 2) and field relationships, OL 7 is probably the best representative of silicate lavas directly related to the natrocarbonatite activity of Oldoinyo Lengai. Under this assumption, the trace element signature of both compositions can be compared.

Natrocarbonatites from Oldoinyo Lengai are alkali-rich, with moderate CaO and insignificant Si, Ti, Al, Fe, Mg contents. P_2O_5 is about 0.8 wt%, and is well below carbonatite averages (Woolley and Kempe 1989). The high contents of the halogens Cl and F, and of SO_2 and CO_2 contrast with the almost anhydrous nature of these rocks. Although natrocarbonatites from Lengai are notoriously high in fluorine (Dawson 1962, 1989; Gittins and McKie 1980; Gittins 1989; Keller and Krafft 1989, 1990; Dawson et al. 1990), the F/Cl ratio is less than 1, unlike most carbonatites, in which F is much more abundant than Cl (Woolley and Kempe 1989). The fractionated nature of natrocarbonatites is emphasized by the very low contents of the compatible elements such as Ni, Cr, Sc, V.

The incompatible trace element patterns show several unusual depletion and enrichment effects compared to the general behavior of incompatible elements in mantle-normalized spidergraphs (Fig. 1) or compared to carbonatites in general (Tables 2, 3). High concentrations of Ba and Sr are a general feature of carbonatites, but the concentrations in natrocarbonatites are distinctly higher than the carbonatite average (SrO 1.4 wt%, BaO 1.4 to 2.4 wt%, Keller and Krafft 1990). Ba/Sr ratios in carbonatites are generally below one, whereas natrocarbonatites show a ratio of 1.0–1.8 that increases with fractional crystallization involving nyerereite and gregoryite. Distinct negative anomalies are shown, in particular, by Ta, Zr, Hf, Y and the HREE. Nb is usually enriched and one of the most distinctive elements in carbonatites. The range of 23–36 ppm Nb in porphyritic natrocarbonatite (Keller and Krafft 1990) is anomalously low and requires explanation, as no Nb fractionating phases such as pyrochlore or niobian perovskite are known at Lengai. All of the silicate rocks of Oldoinyo Lengai show higher Nb than the natrocarbonatites (about ten times, see OL 7 in Table 2, and Donaldson et al. 1987). Ta is even more depleted than Nb. Extreme Nb/Ta ratios of >1000 in natrocarbonatites compared to a value of 61 in the associated combeite nephelinite indicate that Nb and Ta are geochemically decoupled at this stage (Fig. 2). Low Ta in natrocarbonatites can be explained on the basis of the experimentally determined distribution coefficients of Hamilton et al. (1989). However, no such data presently exist for Nb.

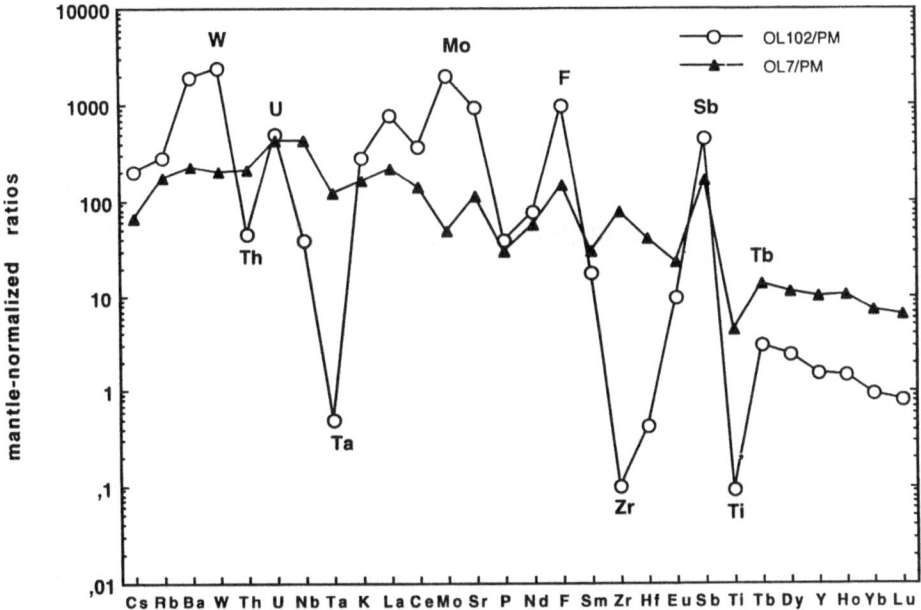

Fig. 1. Trace element abundances in natrocarbonatite (OL 102) and wollastonite-combeite nephelinite (OL 7) normalized to primitive mantle concentrations. (PM values from Sun and McDonough 1989)

Th and U are both enriched in carbonatites compared to most igneous rocks. Woolley and Kempe (1989) report averages for calciocarbonatites of 52 ppm for Th and 8.7 ppm for U. In natrocarbonatite, Th and U exhibit a divergent behaviour, with lower concentrations for Th (3.77 ppm) than for U (10.6 ppm). During fractionation of the natrocarbonatite phenocryst phases, Th and U behave as strongly incompatible trace elements, and the residual melt is enriched in both elements. The distinctively low Th/U ratio is 0.36 and 0.33 for the two natrocarbonatite varieties (Table 2). Th/U ratios in carbonatites, along with most igneous rocks, are usually >1 (Woolley and Kempe 1989). Our comparative data for calciocarbonatites (Table 3) show Th/U ratios of ca. 3, similar to most mantle-derived magmas. Sun and McDonough (1989) list data for chondrites, primitive mantle and oceanic basalts, and all have Th/U ratios of 2.5–4.0. Also the silicate magmas of Oldoinyo Lengai have Th/U ≥2 (Table 2; Dawson and Gale 1970; Donaldson et al. 1987), contrasting sharply with the natrocarbonatite Th/U of 0.35. Similarly low Th/U ratios have been reported for the Lengai lavas of the 1960 eruption (Dawson and Gale 1970; Williams et al. 1986). Given the similar Th/U ratios for the natrocarbonatite and its phenocrysts, nyerereite and gregoryite (Table 2), fractional crystallization will not significantly affect the Th/U ratio. Fractionation of pyrochlores with their variable Th/U ratios can effectively change the Th/U ratios in carbonatites (Deans 1966; Hogarth 1989; Mariano 1989), but there is no evidence for pyrochlore fractiona-

Table 3. Chemical comparison of representative natrocarbonatite composition OL 102 with calciocarbonatites, dolomitic carbonatite and ferrocarbonatite

No.	Na-Carbonatite O.Lengai OL 102	ø Sövite Woolley and Kempe 1989	Sövite Kaiserstuhl MFCA	Extr. Ca-Carbonatites Kerimasi TK 12	TK 24	Mg-Carb. Iron Hill IH 1.2	Fe-Carb. Homa Bay HB 63
SiO_2	16	2.72	1.00	0.38	0.10	0.47	0.57
TiO_2	0.02	0.15	0.02	bd	bd	bd	0.28
Al_2O_3	bd	1.06	0.44	0.02	0.02	0.02	0.10
Fe_2O_3	0.05	2.25	3.73	0.18	0.08	2.13	10.81
FeO	0.23	1.01	0.65	0.12	0.15	2.94	0.70
MnO	0.38	0.52	0.60	0.44	0.17	0.97	3.80
MgO	0.38	1.80	2.25	1.81	0.25	19.50	0.35
Cao	14.02	49.12	47.70	51.46	54.07	32.25	43.52
SrO	1.42	0.86	1.06	1.11	0.55	0.51	0.19
BaO	1.66	0.34	0.15	0.31	0.07	0.05	0.88
Na_2O	32.22	0.29	0.11	0.29	0.16	0.04	0.01
K_2O	8.38	0.26	0.05	0.03	0.02	0.01	0.03
P_2O_5	0.85	2.10	2.65	1.74	1.90	0.86	0.33
CO_2	31.55	36.64	37.18	40.58	40.59	40.92	36.07
Cl	3.40	0.08	0.07	0.20	0.05	0.43	0.09
F	2.50	0.29	0.19	0.36	0.34	0.41	0.04
SO_3	3.72	0.88	0.13	0.47	0.55	0.01	0.05
H_2O^+	0.56	0.76	0.33	0.30	0.80	0.12	1.71
$-O=F,Cl$	-1.82	-0.14	-0.01	-0.15	-0.15	-0.27	-0.03
Total	99.68	100.99	98.31	99.65	99.72	00.52	99.50
ppm							
V	116	80	180	115	88	6	123
Sc	bd	7	1.7	0.02	0.11	9.46	5.9
Cr	bd	13	<5	12	11.2	28	20
Ni	bd	18	<5	bd	bd	bd	bd
Ba	14300	3045	1360	2798	627	442	7906
Sr	11975	7272	8967	9423	4640	4292	1608
Rb	178	14	2.4	bd	bd	bd	<10

Cs	6.4	0.12	bd	bd	bd	bd
Zr	bd	45	24	23	25	49
Nb	28	4500	121	27	461	584
Zn	88	290	213	130	61	1480
Ga	<20	6.2	<0.5	<0.5	<0.5	<1.5
As	18.0	4.4	3.5	5.14	0.35	6.0
Se	<0.5	<1	bd	bd	bd	bd
Br	68.7	1.1	0.79	2.83	2.83	bd
Mo	125	na	<1	<1	<1	13
W	48.9	<0.5	0.78	0.95	0.95	<0.5
Sb	2.2	0.17	0.58	0.06	0.06	0.17
Hf	0.13	0.39	<0.1	<0.1	<0.1	<0.3
Ta	<0.03	5.11	0.22	4.27	4.27	<0.04
Th	3.77	7.73	0.81	5.64	5.64	61.6
U	10.6	2.59	2.96	0.85	2.38	1.1
La	545	605	228	141	143	1062
Ce	645	1460	223	203	322	1974
Pr	–	125	17	11	40.2	210
Nd	102	383	67.3	39.7	138	750
Sm	7.8	48	9.99	6.05	17	96.8
Eu	1.62	13.3	3.03	2.04	4.02	24.6
Tb	0.32	3.98	1.08	0.88	0.90	7.0
Dy	1.75	19.8	5.48	4.67	3.06	33
Yb	0.46	6.3	2.14	2.93	0.44	7.98
Lu	0.058	0.83	0.31	0.43	0.0444	1.06
Y	7	95	32	34	8	97
Ba/Sr	1.2	0.15	0.29	0.14	0.10	4.9
La/Yb	1129	96	106.5	48.1	325	133.1
La$_N$/Sm$_N$	42.6	7.7	14.35	14.65	5.29	6.9
Th/U	0.36	2.98	0.27	0.19	0.68	56
Th/Ta	>150	1.5	3.68	0.80	46.2	>1540

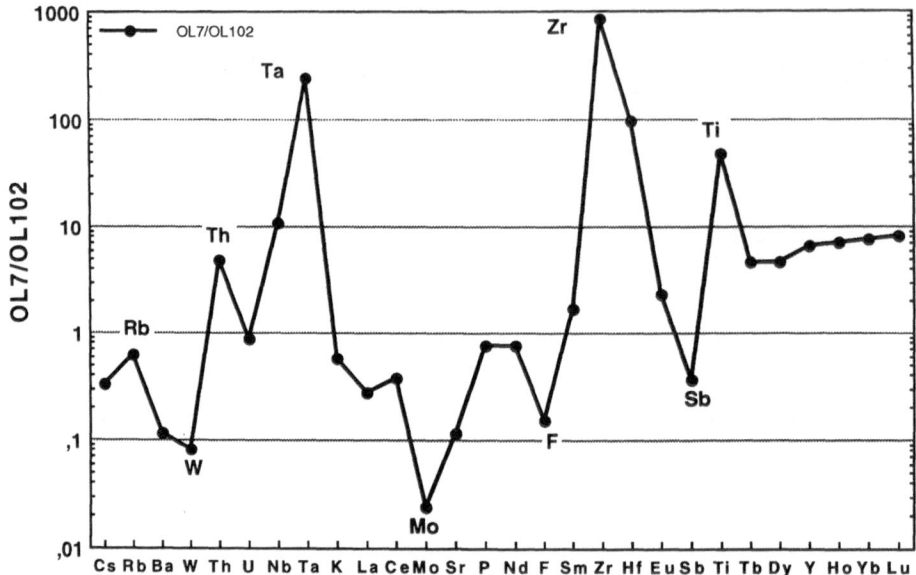

Fig. 2. Trace element ratios for combeite nephelinite (*OL 7*)/natrocarbonatite (*OL 102*) showing the relative enrichment and depletion in both compositions

tion in the evolution of Oldoinyo Lengai. Thus, low Th/U, resulting basically from Th depletion, is a further distinctive geochemical signature of Lengai natrocarbonatites which must be explained in any petrogenetic model. Strong preference of Th for silicate melts has been demonstrated by Hamilton et al. (1989). As with Th/U, different degrees of depletion in generally geochemical coherent element pairs have led to anomalous ratios for Zr/Nb, Zr/Hf, and Nb/Ta, and these ratios are considered distinctive of natrocarbonatites.

Exceptional trace element enrichment relative to most other igneous rocks, including carbonatites, is shown by Sr, Ba, K, Rb, Cl, F, Mo, W, As, Sb, and Br. Concentrations for Mo (125–188 ppm) and W (49–77.5 ppm) in Oldoinyo Lengai natrocarbonatites (Table 2) are among the highest ever reported in non-mineralized magmatic rocks. Mo and W do not exceed a few ppm in most igneous rocks but only a few Mo and W values are available for carbonatites (Nelson et al. 1988). The maximum Mo content is 18 ppm from 18 analyzed carbonatites. Woolley and Kempe (1989) do not report Mo values for any calciocarbonatite, but list three values (26–94 ppm) for ferrocarbonatites. Deans (1966) refers to molybdenum enrichment of up to 0.2% MoO_3 in lateritic soils overlying certain carbonatites. Enrichment of Mo, along with Cl and F, could suggest a certain mobility as volatile complexes, but the Mo/W ratios of the Lengai samples (2.55–2.43) compare well with magmatic mantle systems. Sun and McDonough (1989) give an average ratio of 3.15 for primitive mantle. Compared to primitive mantle, Mo and W of natrocarbonatites show the highest degree of enrichment (Fig. 1), that is only paralleled by Ba and approached by

Sr, U, La, F, and Sb. For Pb we have no values in our data set, but Pb concentrations of 120 and 118 ppm (Williams et al. 1986; Dawson et al. 1990) reported for porphyritic natrocarbonatite show that Pb has the same degree of enrichment as Mo and W, i.e. 1500–2000 times primordial mantle. Woolley and Kempe (1989) give average Pb values of 56 ppm for calciocarbonatites, 89 ppm for magnesiocarbonatites, and 217 ppm for ferrocarbonatites. However, these averages are based on too few analyses with large variations to be of significance.

3 Rare Earth Elements in Natrocarbonatites and Their Phenocrysts

Rare earth elements (REE) were determined for natrocarbonatite and for the phenocryst phases nyerereite and gregoryite. The data for the carbonatite samples are given in Table 2, together with the REE content in the related combeite-nephelinite rim lava (OL 7). The data are shown graphically in the chondrite-normalized REE diagrams of Figs. 3 and 4.

Natrocarbonatite lavas are characterized by an extremely fractionated REE pattern, resulting from LREE enrichment (up to 1700–2400 times chondrite for

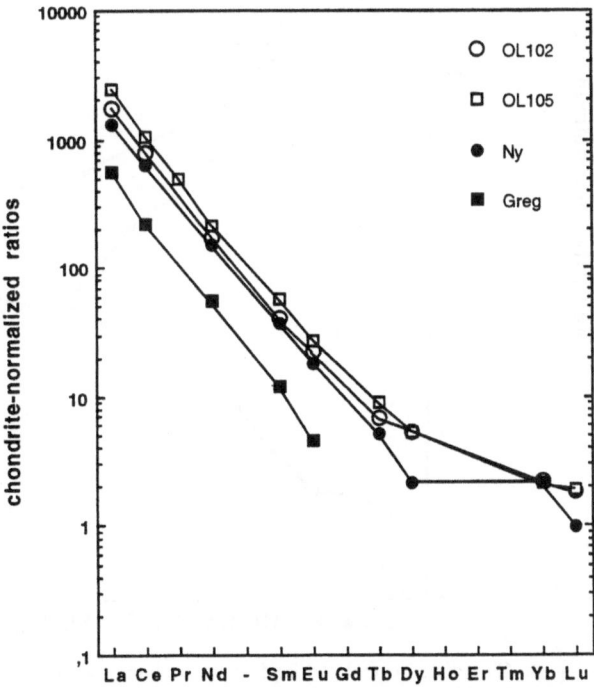

Fig. 3. Chondrite-normalized REE-pattern for whole-rock natrocarbonatites (porphyritic natrocarbonatite *OL 102*, and aphyric natrocarbonatite *OL 105*), and nyerereite and gregoryite phenocrysts. (Normalizing values from Sun and Hanson 1976)

La) and HREE depletion (Yb and Lu values <2 times chondrite). This results in characteristically high LREE/HREE ratios. The slope of LREE enrichment, expressed as (La/Sm)$_N$ ratios of ca. 42–43 (Table 2), is one of the steepest observed in igneous rocks. This steep slope extends to Tb and Dy, and only the pattern for the heaviest REE flattens out (Fig. 3). This REE pattern obtained for natrocarbonatites is highly distinctive compared to other carbonatites (e.g. Eby 1975; Hornig 1988; Nelson et al. 1988; Woolley and Kempe 1989). The distinctive LREE enrichment of natrocarbonatites is not repeated in the REE pattern of the nephelinite OL 7. The cross-over pattern in Fig. 4 shows that its LREE contents are less enriched and the HREE are less depleted than in the natrocarbonatites.

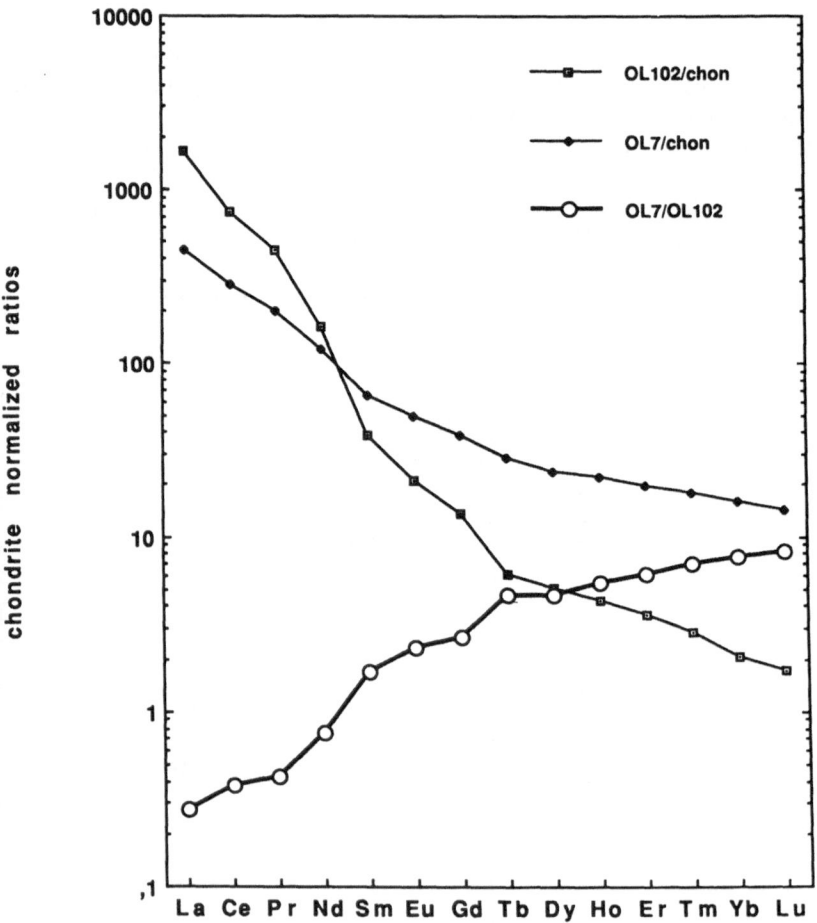

Fig. 4. Chondrite-normalized REE plots showing the cross-over of the REE pattern of nephelinite (*OL 7*) and natrocarbonatite (*OL 102*). Preferential partitioning of LREE into the carbonate melt, and of HREE into a coexisting silicate magma, match the experimental results of Hamilton et al. (1989). The ratio *OL 7/OL 102* approximates the melt/melt K$_D$S between the two supposedly immiscible magmas

Nyerereite and gregoryite have lower REE concentrations than the bulk rock, but their patterns, in the normalized plot of Fig. 2, are perfectly parallel. As a result of nyerereite and gregoryite fractionation the aphyric natrocarbonatite OL 105 is REE-enriched in comparison to the porphyritic composition represented by OL 102 (Table 2). Fractionation of the phenocryst phases has not changed the general shape of the REE pattern. This observation agrees with that of Hornig (1988), who showed that for pairs of calcite/bulk carbonatite and dolomite/bulk carbonatite the relative distribution of REE in carbonatites and in their carbonate phases is usually subparallel, and therefore that the shape of the REE pattern is not affected by the fractionation of crystallizing carbonate phases.

4 Natrocarbonatite Trace Element Signature vs. "Normal" Carbonatites

Despite several specific trace element studies of individual carbonatite occurrences (e.g. Cullers and Medaris 1977; Eby 1975; Gerasimowsky 1978), and compilations by Woolley and Kempe (1989) and Nelson et al. (1988), we are still far from quantitatively understanding the large variation and in some cases divergent behaviour of trace elements in carbonatites. A detailed trace element characterization is available only for a limited number of carbonatites. For comparison with the composition of natrocarbonatite, we have used the sövite average of Woolley and Kempe (1989), and new trace element data for five different carbonatite types (Table 3, Fig. 5). These include a calcite-apatite-magnetite sövite from Kaiserstuhl (MFCA), a rauhaugite (magnesiocarbonatite) from Powderhorn/Iron Hill, Colorado (IH 1.2), and a ferrocarbonatite from Okuge Hill, Homa Bay area, Kenya (HB 63). Of special interest are two extrusive calcitic carbonatites (TK 12, 24) from Kerimasi, the carbonatite volcano next to Oldoinyo Lengai. TK 12 is from a carbonatitic tuff breccia exposed in the walls of the Loolmurwak explosion crater at the northeastern foot of Kerimasi and contains the polycrystalline calcitic pseudomorphs interpreted by Hay (1983) and Deans and Roberts (1984) to represent pseudomorphs of former Na-carbonates (nyerereite). TK 24 represents the carbonatitic lavas from the summit area of Kerimasi which exhibit under cathodoluminescence spectacular primary zoning of calcite phenocrysts (Mariano and Roeder 1983). The models proposed by Hay (1983), Deans and Roberts (1984) and Dawson et al. (1987) invoke loss of alkalies from natrocarbonatites to produce calcitic carbonatites. In this case it should be possible to recognize the trace element signature of natrocarbonatite, at least for those elements which are not readily mobilized. The extrusive calcitic carbonatites TK 12 and TK 24 from Kerimasi do not show any of the distinctive trace element characteristics of natrocarbonatites. Equally, non of the other carbonatite types available for comparison show the spectrum of trace element features characteristic of natrocarbonatite. In particular, the unusually high concentrations of Mo, W, As, Sb, and extreme

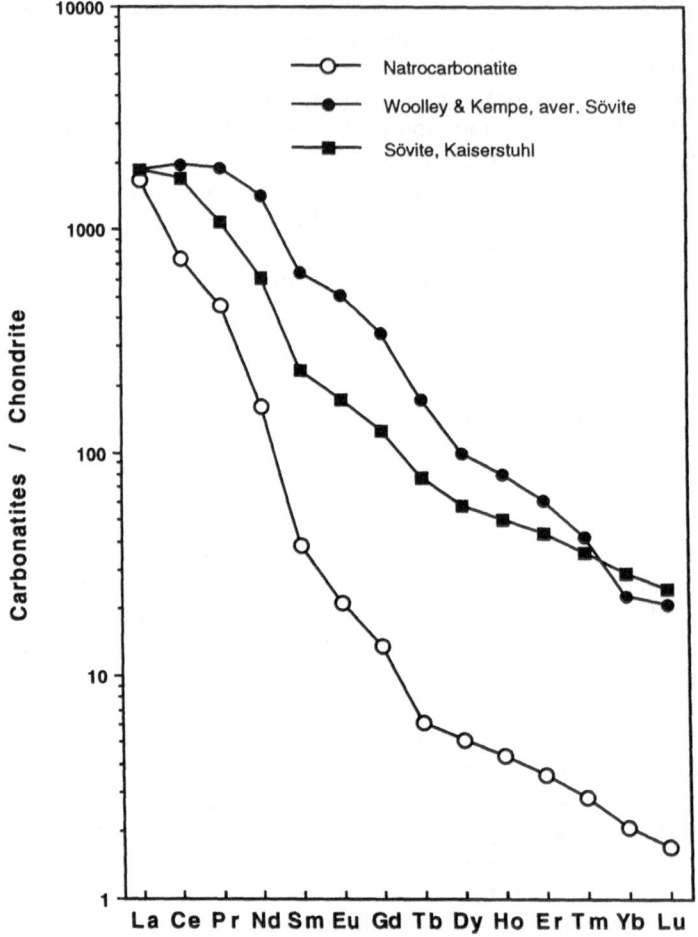

Fig. 5. Comparision of the typical REE pattern in natrocarbonatite with calciocarbonatites (sövites) showing the characteristic steep slope of the LREE/MREE part of the diagram and the unusually low HREE contents, contrasting with the REE behaviour in alkali-poor carbonatites. (Data from Tables 1, 3, and Woolley and Kempe 1989)

LREE/HREE ratios are quite distinct to those found in other "normal" carbonatites (Fig. 5).

The trace element signature of natrocarbonatites shows a number of features which are not readily explained by crystal/liquid fractionation related to the associated silicate lavas. These features include: (1) The negative anomalies of Ta, Zr, Hf, Ti, and to a lesser extent Th and Nb, in mantle-normalized spidergrams (Fig. 1). (2) The positive anomalies for Ba, Sr, Mo, W, F, Sb and the LREE, with an enrichment, relative to primordial mantle (Sun and McDonough 1989), up to 2500. (3) The decoupling of geochemically related element pairs, to an extent which is unlikely to occur during partial melting processes, or fractional crystallization models starting from the evolved silicate lavas of Oldoinyo Lengai.

Hamilton et al. (1989) have experimentally determined the distribution coefficients (K_Ds) of the trace elements Ba, Ta, Zr, Hf, Cu, Cr and the REEs (La, Ce, Sm, Eu, Gd, Yb, Lu) between silicate melts (representing the phonolites and nephelinites of Oldoinyo Lengai) and alkaline carbonatite compositions. Experiments were carried out in the pressure range 1–6 kb and at temperatures ranging from 1050–1250 °C. Although many factors have yet to be considered, in particular how the data can be extrapolated to lower temperatures, some general conclusions can be drawn from these experiments:

1. Ta and Hf can be expected to be strongly enriched in the silicate melt, and a conjugate immiscible alkalic carbonatite would be extremely depleted in these elements. Thus, the extremely low Ta and Hf contents of natrocarbonatites, compared to associated silicate lavas, are consistent with liquid immiscibility. The K_DS silicate/carbonatite for these elements appear even to dramatically increase at temperatures lower than the 1050 °C used in the experiments of Hamilton et al. (1989).

2. Ba and Sr are strongly enriched in an immiscible alkali carbonate melt in equilibrium with phonolite or nephelinite. For Ba the effect seems stronger if extrapolated to lower temperatures (Hamilton et al. 1989, Figs. 16.5 and 16.10). This can easily explain the extreme Ba- and Sr- enrichment and high Ba/Sr in natrocarbonatites compared to the evolved nephelinite (Table 1).

3. Zr exhibits a pronounced negative anomaly in the normalized abundance pattern of natrocarbonatite (contents <10 ppm) but high concentration (800 ppm) in the nephelinite OL 7. This again is in general agreement with the experimental results of Hamilton et al. (1989) showing preferential partitioning of Zr into the silicate melt.

4. An important result obtained by Hamilton et al. (1989) for the REE distribution between silicate and carbonatite melts is the consistently lower K_D's for LREE compared to HREE. This must result in preferential enrichment of La, Ce, etc into a carbonatite melt in equilibrium with an undersaturated silicate melt, and conversely, in depletion of the heavy REE, as shown by the REE pattern in natrocarbonatites (Fig. 4).

5 Conclusions

The natrocarbonatites of Oldoinyo Lengai have a characteristic trace element signature. They are characterized by extreme LREE/HREE enrichment and have almost the highest La/Sm and La/Yb ratios reported of any igneous rocks. The trace element signature of natrocarbonatites is further characterized by very high concentrations of Sr, Ba, Rb, K, light and middle REE, Mo, W, U, As, Sb and strong depletion in Nb, Ta, Hf, Zr, Ti, Y, Yb and Lu. Allegedly calcified natrocarbonatites from other volcanoes (Deans and Roberts 1984; Dawson et al.

1987; Keller 1989) should have retained some of the characteristics of the natrocarbonatite signature. Extrusive carbonatites from Kerimasi do not show this linkage.

On the basis of available experimental data, the unusual relative enrichment or depletion of otherwise closely correlated incompatible trace elements can best be explained by the partitioning of these elements between immiscible silicate and carbonatite melts. A model of liquid immiscible separation of natrocarbonatite from peralkaline nephelinites of Oldoinyo Lengai best explains the trace element data, and is consistent with the volcanological and petrological observations.

6 Appendix: Analytical Methods

Major elements and the trace elements V, Y, Zr, Nb, were determined by XRF on a Philips PW 1450 sequential spectrometer at the Mineralogisch-Petrographisches Institut in Freiburg, using Li-tetraborate fuses and powder pellets. XRFA uses a special carbonatite program with international standards and oxide mixes for calibration. REE and the rest of the trace elements were determined by INAA at the Max-Planck-Institut in Mainz. Cl and F were analyzed by ion-selective electrodes, and CO_2 and SO_3 by colorometric titration using a Coulomat equipment.

Acknowledgements. This research project was supported by DFG grant Ke 136/22-1. Helpful suggestions on an early draft of the manuscript by Gerhard Brey and Ingrid Hornig-Kjarsgaard, and the thorough reviews by Keith Bell, Albrecht Hofmann, Gail Mahood and by an anonymous reviewer are gratefully acknowledged.

References

Bell K, Blenkinsop J (1989) Neodymium and strontium isotope geochemistry of carbonatites. In: Bell K (ed) Carbonatites – genesis and evolution. Unwin Hyman, London, pp 278–300
Cullers RL, Medaris G Jr (1977) Rare earth elements in carbonatite and cogenetic alkaline rocks. Examples from Seabrook Lake, Callander Bay, Ontario. Contrib Mineral Petrol 65:143–153
Dawson JB (1962) Sodium carbonate lavas from Oldoinyo Lengai, Tanganyika. Nature 195: 1075–1076
Dawson JB, Gale NH (1970) Uranium and thorium in alkalic rocks from the active volcano Oldoinyo Lengai (Tanzania). Chem Geol 6:221–231
Dawson JB, Garson MS, Roberts B (1987) Altered former alkalic carbonatite lava from Oldoinyo Lengai, Tanzania: inferences for calcite carbonatite lavas. Geology 15:765–768
Dawson JB, Smith JV, Steele IM (1989) Combeite ($Na_{2.33}Ca_{1.74}others_{0.12}$) Si_3O_9 from Oldoinyo Lengai, Tanzania. J Geol 97:365–372
Dawson JB, Pinkerton H, Norton GE, Pyle DM (1990) Physicochemical properties of alkali carbonatite lavas: data from the 1988 eruption of Oldoinyo Lengai. Geology 18:260–263
Deans T (1966) Economic mineralogy of African carbonatites. In: Tuttle OF, Gittins J (eds) Carbonatites. Wiley, New York, pp 385–413

Deans T, Roberts B (1984) Carbonatite tuffs and lava clasts of the Tinderet foothills, western Kenya: a study of calcified natrocarbonatites. J Geol Soc Lond 141:563–580

Donaldson CH, Dawson JB, Kanaris-Sotiriou R, Batchelor RA, Walsh JN (1987) The silicate lavas of Oldoinyo Lengai, Tanzania. Neues Jahrb Miner Abh 156:247–279

Eby GN (1975) Abundance and distribution of the rare earth elements and yttrium in the rocks and minerals of the Oka carbonatite complex, Quebec. Geochim Cosmochim Acta 39:597–620

Freestone IC, Hamilton DL (1980) The role of liquid immiscibility in the genesis of carbonatites – an experimental study. Contrib Mineral Petrol 73:105–117

Gerasimovsky VI (1978) Geochemistry of the carbonatites of the East African Rift Zones. In: Proc 1st Int Symp on Carbonatites, Pocos de Caldas, Minas Gerais, Brasil. Ministerio das Minas e Energia, Brasilia, pp 207–212

Gittins J (1989) The origin and evolution of carbonatite magmas. In: Bell K (ed) Carbonatites – genesis and evolution. Unwin Hyman, London, pp 580–600

Gittins J, McKie D (1980) Alkalic carbonatite magmas: Oldoinyo Lengai and its wider applicability. Lithos 13:213–215

Hamilton DL, Bedson P, Esson J (1989) The behaviour of trace elements in the evolution of carbonatites. In: Bell K (ed) Carbonatites – genesis and evolution. Unwin Hyman, London, pp 405–427

Hay RL (1983) Natrocarbonatite tephra of Kerimasi volcano, Tanzania. Geology 11:599–602

Hogarth DD (1989) Pyrochlore, apatite and amphibole: distinctive minerals in carbonatite. In: Bell K (ed) Carbonatites – genesis and evolution. Unwin Hyman, London, pp 105–148

Hornig I (1988) Spurenelementuntersuchungen an Karbonatiten mit Hilfe der ICP-Atomemissionsspektroskopie. PhD Thesis, University Freiburg, 273pp

Javoy M, Pineau F, Staudacher T, Cheminée JL, Krafft M (1989) Mantle volatiles sampled from a continental rift: the 1988 eruption of Oldoinyo Lengai. TERRA Abstr 1:324

Keller J (1989) Extrusive carbonatites and their significance. In: Bell K (ed) Carbonatites – genesis and evolution. Unwin Hyman, London, pp 70–88

Keller J, Krafft M (1989) Composition of natrocarbonatite lavas, Oldoinyo Lengai 1988. TERRA Abstr 1:286

Keller J, Krafft M (1990) Effusive natrocarbonatite activity of Oldoinyo Lengai, June 1988. Bull Volcanol 52:629–645

Kjarsgaard BA, Hamilton DL (1989) The genesis of carbonatites by immiscibility. In: Bell K (ed) Carbonatites – genesis and evolution. Unwin Hyman, London, pp 388–404

Koster van Groos AF, Wyllie PJ (1968) Liquid immiscibility in the join $NaAlSi_3O_8$-Na_2CO_3-H_2O and its bearing on the genesis of carbonatites. Am J Sci 266:932–967

Le Bas MJ (1987) Nephelinites and carbonatites. In: Fitton JG, Upton BGJ (eds) Alkaline igneous rocks. Geol Soc Spec Publ Lond 30:53–83

Mariano AN (1989) Nature of economic mineralization in carbonatites and related rocks. In: Bell K (ed) Carbonatites – genesis and evolution. Unwin Hyman, London, pp 149–176

Mariano AN, Roeder PL (1989) Kerimasi – a neglected carbonatite volcano. J Geol 91:449–455

Nelson DR, Chivas AR, Chappell BW, McCulloch MT (1988) Geochemical and isotopic systematics in carbonatites and implications for the evolution of ocean-island sources. Geochim Cosmochim Acta 52:1–17

Peterson TD (1989a) Peralkaline nephelinites I. Comparative petrology of Shombole and Oldoinyo Lengai, East Africa. Contrib Mineral Petrol 101:458–478

Peterson TD (1989b) Peralkaline nephelinites II. Low pressure fractionation and the hypersodic lavas of Oldoinyo Lengai. Contrib Mineral Petrol 102:336–346

Peterson TD (1990) Petrology and genesis of natrocarbonatite. Contrib Mineral Petrol 105:143–155

Sun SS, Hanson GN (1976) Rare earth element evidence for differentiation of McMurdo volcanics, Ross Island, Antarctica. Contrib Mineral Petrol 54:139–155

Sun SS, McDonough WF (1989) Chemical and isotopic systematics of oceanic basalts: implications for mantle composition and processes. In: Saunders AD, Norry MJ (eds) Magmatism in the oceanic basins. Geol Soc Spec Publ Lond 42:313–345

Twyman JD, Gittins J (1987) Alkalic carbonatite magmas: parental or derivative? In: Fitton JG, Upton BGJ (eds) Alkaline igneous rocks. Geol Soc Spec Publ Lond 30:85–94

Wallace ME, Green DH (1988) An experimental determination of primary carbonatite magma composition. Nature 335:343–346

Williams RW, Gill JB, Bruland KW (1986) Ra-Th disequilibria systematics: Timescale of carbonatite magma formation at Oldoinyo Lengai volcano, Tanzania. Geochim Cosmochim Acta 50:1249–1259

Woolley AR, Kempe DRC (1989) Carbonatites: nomenclature, average chemical compositions, and element distribution. In: Bell K (ed) Carbonatites – genesis and evolution. Unwin Hyman, London, pp 1–14

Wyllie PJ, Baker MB, White BS (1990) Experimental boundaries for the origin and evolution of carbonatites. Lithos 26:3–19

Cathodoluminescence Observations of Natrocarbonatites and Related Peralkaline Nephelinites at Oldoinyo Lengai

U. KOBERSKI and J. KELLER

Abstract

Cathodoluminescence (CL) is utilized for the identification and characterization of mineral phases crystallizing from natrocarbonatites of the recent eruptions of Oldoinyo Lengai. The Na-carbonates, nyerereite and gregoryite, show dominantly Mn-activated orange luminescence, quite similar to calcites and dolomites of carbonatites. Nyerereite from older carbonatites is distinguished by atypical blue luminescence. CL also reveals the presence and textural arrangement of submicroscopic sylvite, fluorite and apatite. Among the alteration products in natro-carbonatites, nahcolite is characterized by green CL colours whereas pirssonite and gaylussite luminesce red and blue.

Zoning in nepheline, wollastonite, combeite and apatite from a combeite-nephelinite lava is documented. The CL spectroscopy shows that Mn^{2+} is the main activator in wollastonite, combeite, and in the green cores of apatite. The spectral analysis of blue apatite indicates REE activation, considered typical of apatites from carbonatites.

1 Introduction

The special geochemistry and mineral chemistry of carbonatites makes these rocks particularly promising for cathodoluminescence investigations. Chemical features favourable for CL studies are the presence of activator elements such as REE and Mn, and low concentrations of CL-quenching elements such as Fe. Pioneering CL studies of carbonatites have been presented by Mariano (1978, 1988, 1989) and Mariano and Roeder (1983). Mariano and Roeder (1983) have applied CL to extrusive carbonatites of Kerimasi and have shown that CL can reveal primary zoning features in calcite phenocrysts not detectable by other methods. Similar results have been obtained by Keller (1989) for zoned por-phyritic calcite in carbonatite extrusives from the Kaiserstuhl Alkaline Volcanic Complex, Germany. Other carbonatite minerals that have been studied with CL are apatite, dolomite, fluorite, monticellite, wollastonite and several REE-bearing minerals (Mariano 1989). Complex growth structures in crystals, internal zonation, intergrowths of different mineral phases, minor constituents of the groundmass and the identification of different generations of the same mineral can often be easily detected with the help of CL. No cathodoluminescence data

Institut für Mineralogie, Petrologie und Geochemie, Universität Freiburg, Albertstr. 23b, 79104 Freiburg, Germany

exist so far for natrocarbonatites. It is shown that CL is a powerful tool for the characterization and recognition of the specific minerals of natrocarbonatites and their alteration products.

2 Material and Technique

Cathodoluminescence was used to investigate the nature of the phenocrysts and groundmass minerals of natrocarbonatite from the June 1988 eruptive activity of Oldoinyo Lengai (Keller and Krafft 1989, 1990) and comparison was made with a sample from the 1960 eruption. The alteration of natrocarbonatite and related CL effects are described for a pre-1960 nyerereite carbonatite flow and for altered material of the 1988 activity. The cathodoluminescence of a peralkaline combeite-bearing nephelinite spatially related to the recent natrocarbonatite activity of Oldoinyo Lengai (Keller and Krafft 1990) is discussed in connection with the carbonatitic samples.

Cathodoluminescence investigations were carried out using polished thin sections in a vacuum chamber of a cold cathode electron gun (Technosyn Model 8200 MK II) mounted on a polarizing microscope. Under the high-energy electron beam, electrons of certain activators (e.g. REE, transition metals) are excited and emit, by return to the ground state, radiation of distinct energy registered here in the visible part (400–750 nm) of the electromagnetic spectrum. In addition, intrinsic CL can result from structural imperfections or nonstoichiometry (Marshall 1988). Thus, the CL emission and its spectral analysis supply information about ionic substitution, impurities incorporated in the crystals and about imperfections of the crystal structure and lattice defects (Nakagawa et al. 1988). For the spectral analysis of the luminescence, the CL microscope is connected via a liquid-light guide to a high-resolution spectrometer

──▶

Fig. 1. Natrocarbonatite OL 107 (1988). Cathodoluminescence of brown luminescent gregoryite (*Gr*) showing orange exsolutions of nyerereite. Orange nyerereite (*Ny*) occur as phenocrysts and needle-shaped microphenocrysts. Ca-silicate (*arrow*) luminesces with light blue colour, while sylvite of the groundmass shows turqoise blue colour. Minute intergrowths of fluorite, sylvite, nyerereite and gregoryite in the matrix show dull blue colours grading into dark orange with increasing Na-carbonate contents. CL conditions: Technosyn, 19 kV, 0.49 mA. Photographic conditions: 155 s, ASA 400. Width of photograph: 2 mm
Fig. 2. Natrocarbonatite OL 1960 (1960). Arborescent sylvite is conspicuous showing a turquoise blue colour (*arrow*). Dark blue cube near the sylvite aggregate is fluorite. Sub-microscopic sheaf-like intergrowths of fluorite, and sylvite together with Na-carbonates in the matrix luminesce from blue to orange. CL: 21 kV, 0.42 mA. Photo: 324 s, ASA 200. Width of photograph: 2 mm
Fig. 3. Altered natrocarbonatite OL 115 (1988). The start of hydration of nyerereite (*Ny*) is marked by red gaylussite (*white arrow*). Another alteration product is nahcolite with green CL colours (*black arrow*). CL: 19 kV, 0.45 mA. Photo: 263 s, ASA 200. Width of photograph: 2 mm
Fig. 4. Pre-1960 natrocarbonatite OL 4. Primary nyerereite (*Ny*) shows a bright blue CL colour, whereas the hydration products of nyerereite show distinctive red and dark blue colours. CL: 21 kV, 0.4 mA. Photo: 46 s, ASA 1000. Width of photograph: 2 mm

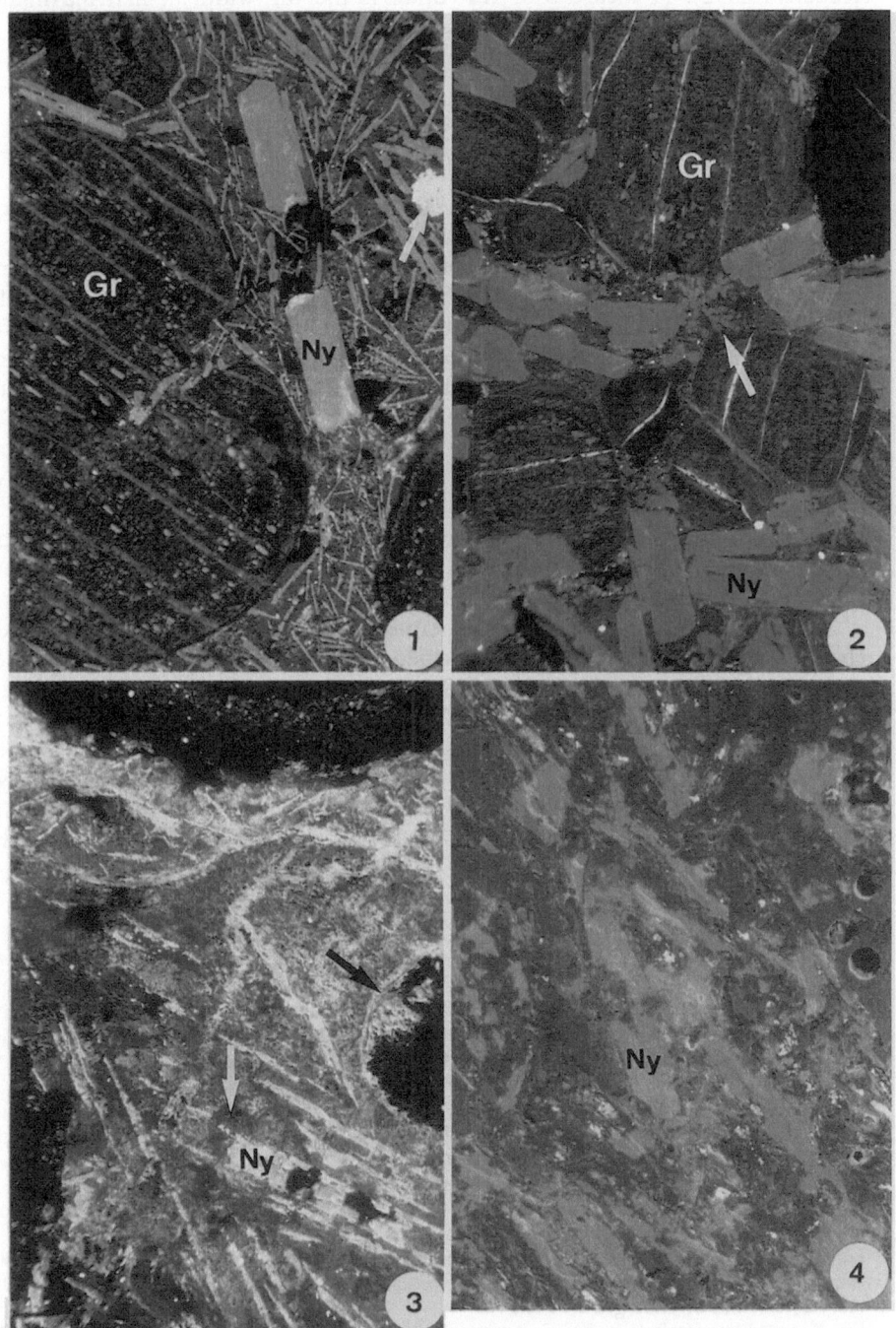

(Koberski 1992). A measure of resolution is the full width at half maximum (FWHM) of a spectral signal, and this is 15 Å.

3 Luminescence of Natrocarbonatite

The samples studied are porphyritic natrocarbonatite lavas with phenocrysts of nyerereite and gregoryite (Keller and Krafft 1990). The matrix consists of microcrystals and quenched phases of the same minerals, in addition to very fine-grained sylvite and fluorite. With the exception of Fe-alabandite, the only ore mineral present in natrocarbonatite (Dawson et al. 1990; Keller and Krafft 1990), all mineral phases show more or less brilliant luminescence induced by their Mn and/or REE contents. Iron contents are very low in nyerereite and gregoryite (Keller and Krafft 1990; Peterson 1990) and therefore the CL quenching effects are negligible. The major luminescence features of mineral phases from the porphyritic natrocarbonatites are shown in Figs. 1–8. Generally, the most important activator in carbonates is Mn^{2+}, which produces a broad band in the orange-red part of the spectrum; the REE can, in some cases, be activators in carbonatitic carbonates (Mariano 1978, 1988) producing narrow spectral bands or peaks.

Nyerereite (Fig. 1, Ny) has the generalized formula $Na_2Ca(CO_3)_2$. Complex chemical substitutions have been described by Gittins and McKie (1980), Keller and Krafft (1990), and Peterson (1990). Keller and Spettel (Chap. 5, this Vol.) provide trace element data including full REE patterns for separated nyerereite and gregoryite. In fresh natrocarbonatite lava from the June 1988 eruption, nyerereite shows, under CL, uniform, bright-orange luminescence (Fig. 1). No chemical inhomogeneity or zones are visible by CL. Needle-shaped nyerereite crystals of the matrix show the same bright orange luminescence as the phenocrysts.

Fig. 5. Pre-1960 natrocarbonatite OL 4. Cores of nepheline xenocrysts (*Ne*) show dull blue luminescence comparable with nepheline phenocrysts of the associated combeite nephelinite (OL 7, Fig. 6). Original nyerereite (*Ny*) shows bright blue luminescence, tabular pirssonite (*white arrow*) luminesces dark blue, and calcite orange (*black arrow*). CL: 20 kV, 0.45 mA. Photo: 204 s, ASA 1000. Width of photograph: 2 mm

Fig. 6. Combeite nephelinite OL 7. Nepheline phenocryst (*Ne*) shows a dull blue, slightly zoned core and non-luminescing rim. Wollastonite inclusions in the rim luminesce bright yellow. Smaller matrix nephelines show no luminescence but contain yellow inclusions (*arrow*). Combeite (*Co*) crystals exhibit light-brown luminescence. Calcite occurs as inclusions in nepheline and in the matrix. Calcite shows orange CL colours. CL: 20 kV, 0.44 mA. Photo: 900 s, ASA 200. Width of photograph: 2 mm

Fig. 7. Combeite nephelinite OL 7. Bright yellow wollastonite (*Wo*) surrounded by light-brown combeite (*Co*). Calcite (*arrow*) luminesces typical orange. CL: 18 kV, 0.52 mA. Photo: 129 s, ASA 200. Width of photograph: 1 mm

Fig. 8. Combeite nephelinite OL 7. Apatite with violet and greenish luminescence shows cyclic zoning and corrosive features. The final stage of apatite crystallization is marked by a light blue homogenous phase. CL: 20 kV, 0.44 mA. Photo: 200 s, ASA 200. Width of photograph: 1 mm

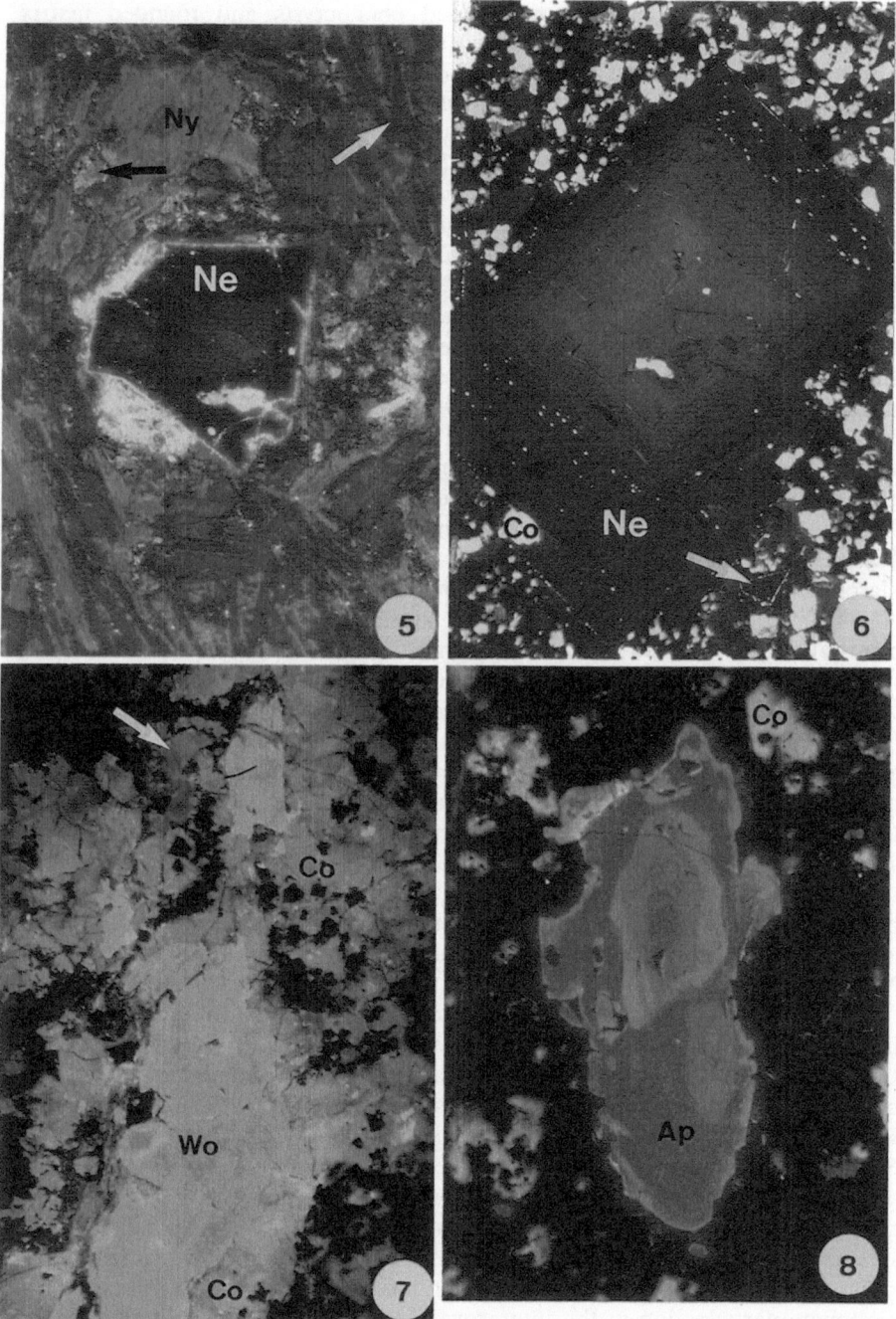

Gregoryite (Gr) occurs as oval-shaped phenocrysts and rounded matrix grains. The generalized composition is [(Na,K)$_2$CO$_3$] and details of chemical substitutions are discussed by Keller and Krafft (1990) and Peterson (1990). Gregoryite shows rather dark brown-orange CL colours (Figs. 1, 2), much darker than nyerereite. CL reveals clearly the presence of nyerereite inclusions (bright orange) in gregoryite phenocrysts (Fig. 1), that appear as exsolved plates oriented parallel to the gregoryite cleavage. The concentric arrangement of exsolved nyerereite points to distinct original zoning in nyerereite solid solution of these gregoryite phenocrysts. The outer rim of gregoryite phenocrysts is generally homogeneous without any exsolution.

Figure 9 presents the emission spectrum of gregoryite in contrast to nyerereite. A typical broad band centred at about 620–630 nm results from Mn^{2+} activation. Mn is also the main CL activator in nyerereite. Mn contents of gregoryite and nyerereite have been determined as 918 and 760 ppm, respectively; Fe contents are 0.04 and 0.02 wt% (Keller and Spettel, Chap. 5, this Vol.). Several sharp peaks appear superimposed upon the Mn band and on the band in the blue region. Examples are peaks at 655, 485, 469, 426 and 389 nm. Peaks at 385, 470 and 487 nm are commonly attributed to Tb^{3+} (e.g. in synthetic fluorite), but only in the presence of the major line at 545 nm (Mariano 1988). Mason and Mariano (1990) show an emission spectrum for synthetic calcite doped with Dy with a narrow peak at 484 nm. Doped calcites show narrow bands at 652 nm for Sm^{3+} and 656 nm for Eu^{3+} (Mason and Mariano 1990). In synthetic fluorite the Sm^{3+} line appears at 650 nm (Mariano 1988). Our spectra show none of the major peaks for elements expected at 545 nm (Tb^{3+}), 602 nm (Sm^{3+}) and 590, 614 nm (Eu^{3+}). These peaks, although with lower intensity, appear also in a "black hole" experiment without any sample. In this experiment the electron beam was used without any sample in the vacuum chamber. In the infrared region of the electromagnetic spectrum (650–899 nm) there are similar-shaped narrow peaks at 782.5, 855 and 871 nm. This experiment suggests that these peaks result from instrumental problems, possibly connected with the bluish purple plasma discharge in the vacuum chamber, typical for cold cathode electron guns. Similar isolated sharp peaks have been observed by Mason and Mariano (1990) in the blue region of the spectrum of calcite doped with Mn.

The occurrence of primary magmatic *sylvite* in natrocarbonatite and the importance of *fluorite* as a matrix phase were first recognized by X-ray diffraction in the 1988 lavas by Keller and Krafft (1989). The mineralogical identification of these very fine-grained phases was achieved with the help of CL. Figures 1 and 2 show characteristic turquoise blue luminescent sylvite grains. Sylvite acquires a purple colour from electron bombardment during CL observations. Fluorite shows blue CL colours, which are different from sylvite because they occur as darker grains without the turquoise colour. Submicroscopic, sheaf-like intergrowths of fluorite together with sylvite, nyerereite and gregoryite are characterized by dull blue colours grading into dark orange with increasing Na-carbonate participation in the intergrowth. Very rare patches of white to light blue luminescent grains (Fig. 1) are present, and these are REE-bearing Ca-silicate grains according to Keller and Krafft (1990). Apatite is visible by its bright blue-violet luminescence. Apatites occur as minute grains ($<20\,\mu$) dispersed in the matrix and included in gregoryite.

Luminescence characteristics of nyerereite and gregoryite in a lava sample from the 1960 eruption (OL 1960) are identical to those found in the 1988 lavas. However, sample OL 1960 does not show the needle-shaped matrix nyerereite (Fig. 2). Arborescent turquoise-blue sylvite and very dark blue fluorite are also present in the 1960 lavas as primary magmatic groundmass phases (Fig. 2, arrow).

Alteration of the 1960 lava is indicated by a green luminescing mineral phase that occurs along cracks in gregoryite phenocrysts. This has been identified as nahcolite ($NaHCO_3$). Nahcolite has been described by Dawson (1962) and Keller and Krafft (1990) as white efflorescences on the surface of recently erupted flows. Alteration of recent natrocarbonatites to earthy pseudomorphs occurs within a few weeks or months after eruption (Keller and Krafft 1990). The alteration of such a sample (OL 115) is marked by the appearance of intense green and red luminescence colours (Fig. 3). The green colours on cracks in gregoryite again reflect nahcolite. Hydration of nyerereite is marked by red-coloured rims around relict nyerereite. This newly formed red phase is probably gaylussite [$Na_2Ca(CO_3)_2 \cdot 5H_2O$], identified by Keller and Krafft (1990) as an important alteration product in recent natrocarbonatite.

A microporphyritic natrocarbonatite lava from the inactive southern part of the crater of Oldoinyo Lengai (OL 4) has been described by Dawson et al. (1987) and attributed to pre-1960 carbonatite activity. The studied samples (e.g. OL 4) are obviously free of gregoryite and have crystallized mainly as nyerereite (Keller and Krafft 1990). Nyerereite occurs as relict cores, rimmed by hydration and alteration zones of pirssonite ($Na_2Ca(CO_3)_2 \cdot 2H_2O$) (Dawson et al. 1987). The hydration products of nyerereite are characterized by either red or dark blue CL colours. It appears that the red colour is typical of hydration rims around nyerereite. The small, tabular, obviously newly formed crystals of pirssonite are mainly dark blue in colour (Figs. 4, 5). The relative proportions of gaylussite and pirssonite in the alteration products of natrocarbonatites is difficult to estimate as gaylussite converts rapidly to pirssonite when the samples are heated to 100 °C during sample preparation.

The blue luminescence of nyerereite in OL 4 contrasts with the orange nyerereite of the lavas from 1960 and 1988 eruptions. The explanation for this differing luminescence is not obvious. No pronounced Mn band is observed in emission spectra of blue nyerereite from the pre-1960 carbonatite flow. Thus, the blue colour could be an intrinsic luminescence resulting from lack of Mn activation, resulting from oxidition of Mn^{2+} to a higher valency state. Blue luminescence is reported in carbonates from Iceland (Marshall 1988) and is attributed to an intrinsic luminescence, that implies structural imperfections. Sippel and Glover (1965) relate blue CL to an unspecified defect in calcite. Such defects could be caused by electron beam damage (Mason and Mariano 1990). On the other hand, Mason and Mariano (1990) reported blue CL colour in synthetic calcite attributed to Eu^{2+} emission. However, in an emission spectra of blue nyerereite the broad band centred at 420 nm, thought to be caused by Eu^{2+}

(Mason and Mariano 1990), is missing, and no indication of anomalous Eu^{2+} contents has been found.

In addition to nyerereite and its hydration products, fluorite and apatite have been identified as magmatic constituents of the pre-1960 natrocarbonatite. Fluorite appears as blue luminescent interstices in the matrix, and apatite occurs as larger grains (up to $\sim120\,\mu$) showing blue-violet luminescence and distinct zoning.

Some secondary calcite, with orange-yellow CL colour, is found as minute grains in the altered matrix and enclosed in nyerereite. Xenocrystic silicate phases commonly appear in the pre-1960 lava and are easily detected by their lack of luminescence. OL 4 contains xenocrysts of sanidine, nepheline, and pyroxene, for which no luminescence was observed. Only wollastonite xenocrysts show a conspicuous, bright yellow luminescence. Some cores of nepheline show dull blue luminescence (Fig. 5), comparable with that shown by nepheline phenocrysts of the associated combeite-nephelinite (Fig. 6).

4 Luminescence of Combeite Nephelinite

The cone of Oldoinyo Lengai is built up dominantly of nephelinitic, phonolitic and carbonatitic lavas (Donaldson et al. 1987; Dawson 1989). Olivine nephelinites and olivine melilitites occur as minor flows from parasitic cones. Peterson (1989) and Keller and Krafft (1990) have discussed a close petrogenetic relationship of wollastonite- and combeite-bearing peralkaline nephelinites with natrocarbonatites (see also Kjarsgaard et al., Chap. 12, this Vol.). A highly peralkaline, combeite-bearing nephelinite is directly associated with the recent natrocarbonatite activity of Oldoinyo Lengai (Keller and Kraft 1990; Keller and Spettel, Chap. 5, this Vol.). The mineral mode is nepheline, combeite, melanite, and pyroxene, with accessory wollastonite, apatite and sphene. Trace element data and the isotopic composition of the combeite nephelinite sample (OL 7) are discussed in Keller and Krafft (1990) and Keller and Spettel (Chap. 5, this Vol.).

Under CL the bright yellow colour of the wollastonite is striking. Wollastonite is a minor constituent occurring as large crystals or glomerophyric clusters surrounded by a reaction rim of combeite crystals (Fig. 7). Wollastonite shows primary zoning, typical of igneous crystallization. Minute inclusions in nepheline phenocrysts are identified as wollastonite by their shape and bright yellow CL colour. The emission spectrum of the yellow wollastonite is characterized by a broad band caused by Mn^{2+} and centred at 561 nm. Marshall (1988) reported a Mn-maximum at 560 nm for pseudo- or α-wollastonite in contrast to 620 nm for β-wollastonite.

Combeite ($Na_{4.3}Ca_{3.7}Si_6O_{18}$) occurs, both as euhedral phenocrysts and as reaction rims around wollastonite. In both cases, combeite shows light yellow-brown luminescence colours (Figs. 6, 7). The rim of combeite phenocrysts commonly shows dark-brown luminescence. In the wollastonite-combeite clusters, a slightly lighter luminescence of combeite adjacent to wollastonite can be

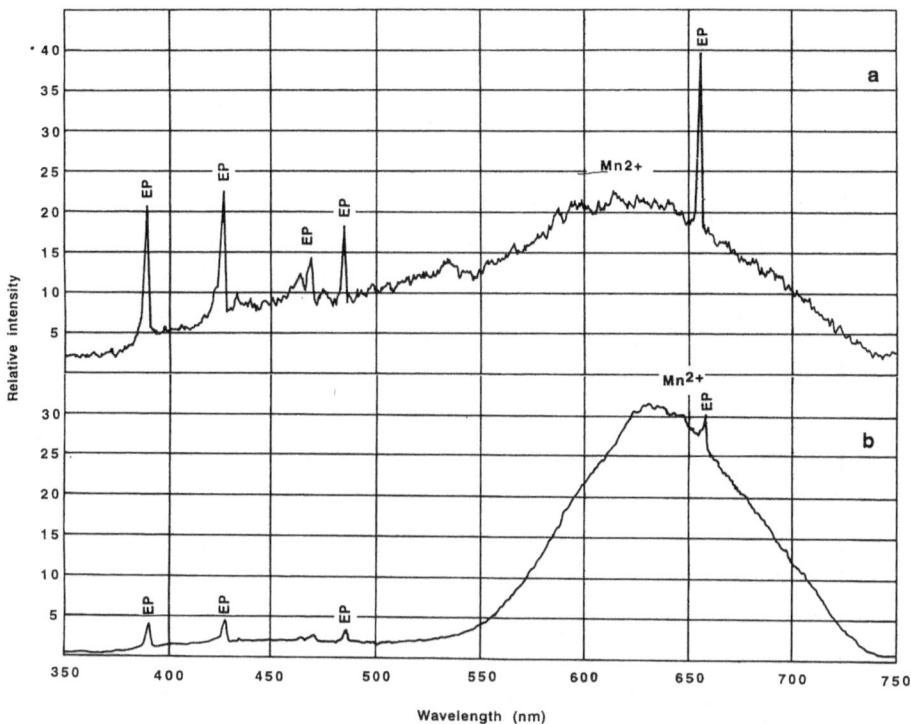

Fig. 9a,b. Natrocarbonatite. CL spectrum of Mn^{2+} activated gregoryite containing 900 ppm Mn and 440 ppm Fe (**a**) and for nyerereite containing 700 ppm Mn and 200 ppm Fe (**b**). Mn^{2+} maxima are centred at 620 nm. The broad band emission in the blue part of the gregoryite spectrum may result from REE activation or may be intrinsic in nature. The CL colour is moderate brown-orange for gregoryite and bright orange for nyerereite. **a** CL conditions: 20 kV, 0.47 mA. Photometer energy: 990 V. **b** CL: 20.5 kV, 0.44 mA. Photometer energy: 950 V. *EP* Electron beam peak

explained by a smaller degree of substitution of the quenching element Fe^{2+} for Ca (Peterson 1989). Luminescence of combeite is caused by Mn^{2+}. The emission maximum of the broad Mn^{2+} band is about 590 nm for combeite.

Nepheline is the major constituent of the nephelinite and shows dull blue, slightly zoned cores with non-luminescing rims (Fig. 6). The outer zones of the nepheline phenocrysts show inclusions or exsolution lamellae with bright yellow colour. These have been identified as wollastonite. The outermost rim of nepheline is marked by a concentration of these inclusions. Smaller nephelines of the matrix do not show any luminescence, but often have zones rich in wollastonite inclusions (Fig. 6, arrow).

Apatite exhibits a complex crystal growth structure (Fig. 8). Cores showing greenish luminescence are enclosed by zones with violet CL. Both apatite phases show cyclic zoning and corrosive contacts between the zones. A light blue homogeneous outer rim of apatite is formed during the last stage of crystallization. Smaller apatites of the matrix do not have green cores. Emission

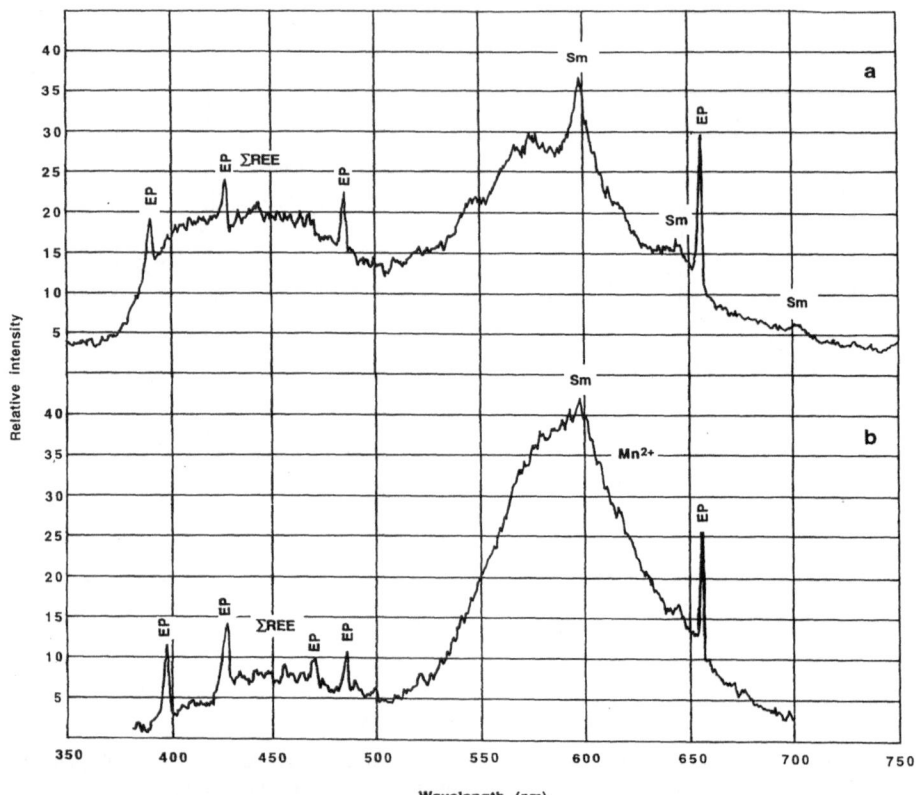

Fig. 10a,b. Combeite nephelinite. **a** Spectrum of a blue-violet apatite from combeite nephelinite with a broad band peaking in the blue region of the spectrum (defect related by the substitution of REE for Ca^{2+}) and a minor broad band in the orange part caused by Mn^{2+}, superimposed by the prominent line of Sm^{3+} at 597 nm. **b** Spectrum of a green apatite core from combeite nephelinite. The emission spectrum shows distinct Mn^{2+} activation with a broad band centred at about 590 nm, a superimposed Sm line at 597 nm and a restrained blue band indicating minor substitution of REE in the apatite structure. **a** CL conditions: 19 kV, 0.47 mA. Photometer energy: 990 V. **b** CL: 19 kV, 0.47 mA. Photometer energy: 990 V. EP = electron beam peak

spectra of a **blue-violet** apatite from OL 7 shows Sm^{3+} activation with the prominent line at 597 nm and a broad band in the blue region (Fig. 10a). Mariano (1988) suggested that a structural defect, caused by the substitution of REE for Ca^{2+}, is mainly responsible for the blue luminescence of carbonatitic apatites. The spectrum of a greenish core shows a more pronounced Mn^{2+} band superimposed by a small line of Sm^{3+} at 597 nm. The blue band is more restrained (Fig. 10b).

Calcite occurs as inclusions in nepheline (Fig. 6), in combeite-wollastonite reaction aggregates (Fig. 7) and in the matrix (Fig. 8). In all cases, calcite luminesces a **typical orange.** Some calcite grains show zonation.

Non-luminescing silicate phases are pyroxene, sphene and melanite as well as some opaques. Peterson (1989) reported sodalite as a minor constituent in the ejected combeite-nephelinite lava blocks in his study from Oldoinyo Lengai. However, sodalite is easily recognizable by the red CL in carbonatites (Mariano 1978), and has not been found in sample OL 7.

5 Discussion and Conclusions

5.1 Natrocarbonatites

Nyerereite and gregoryite, the major phenocryst phases of natrocarbonatite, exhibit a distinct luminescence signature. The spectra of Na-carbonates indicate Mn^{2+} activation with a broad band at about 620 nm, which contrasts with hydrothermal calcite with a band maximum centred at 590 nm. Continuation of the broad Mn band into the blue region of the spectrum points to the influences of REE, as well as to possible effects of intrinsic origin. The Mn band in natural calcite is characterized by a steep slope from 590 nm down to 500 nm. However, the Mn/Fe ratios of Na-carbonates of our study and hydrothermal calcites are about the same. The characteristic emission spectra can be attributed to significant differences in REE pattern and/or to structural differences in the carbonates.

Nyerereite relics from the pre-1960 eruption show atypical blue luminescence colour. In the spectrum of blue nyerereite Mn^{2+} activation is not evident. This is most likely a result of Mn^{2+} being oxidized to a higher valence state, or alternatively no Mn being present in the sample. This is typical of carbonates and phosphates that crystallize or are altered by oxidation in the zone of weathering (Mariano 1978, 1988). Incorporation of Mn in carbonates is influenced by the redox conditions of the system (Carpenter and Oglesby 1976). Total concentration of REE, Mn and Fe and the structural state of the carbonates may also control the luminescence. Therefore the blue colour of pre-1960 nyerereite points to different crystallization conditions of these lavas compared to the recent lavas.

Important magmatic groundmass phases, sylvite and fluorite, are discernible by their luminescence colours and their behaviour after electron bombardment. Teegarden (1966) found that pure sylvite luminesces only weakly, if at all. Luminescence of sylvites from the carbonatite rocks is quite strong. Sylvite can accept many activators (Marshall 1988), including the REE. No spectra have been obtained for sylvite because of its small size.

Fluorite is also a mineral that can accept many activators, including the REE (Marfunin 1979), which generally produce blue colours (Marshall 1988). Dickson (1980) found that fluorite developed a semipermanent purple colour after prolonged bombardment with high-energy electrons and suggested this property as a means of identification of small fluorite crystals. However, in the examined natrocarbonatites only sylvite shows this property.

The examples cited in this chapter clearly show how CL can help resolve the nature of the matrix and help in the identification of minor phases, e.g. microcrystals of apatite, calcite and wollastonite. In addition, the degree of alteration may be estimated with the help of CL due to the appearance of green, red or blue luminescence colours in carbonates. Pirssonite shows either red, blue or brown colours. The gradation of luminescence could reflect different stages of hydration, depending on the temperature of formation. Decreasing temperature increases the lattice order and prevents the admission of activators. Meyers (1974) pointed out the absence of CL in vadose cements of limestones. Therefore red-coloured pirssonite could represent hydration in a post-magmatic state. Brown- and dull blue-coloured pirssonite indicate secondary formation under lower temperatures.

5.2 Combeite-Nephelinite

CL spectroscopy provides further information about the role of activators in mineral phases of silicate rocks related to carbonatitic activity. CL emission of bright-yellow wollastonite and of light brown combeite is caused by Mn^{2+}. The band maximum of wollastonite is about 561 nm, whereas the maximum for combeite is centred at 590 nm. The spectral maximum of CL emission of Mn^{2+} depends on the crystal field and the crystallographic position (Marshall 1988). This accounts for the shift of the Mn^{2+} band maximum in combeite compared to wollastonite. Zoning in euhedral combeite phenocrysts could be attributed to varying substitution of Mn and Fe in the combeite structure.

Zoning and inclusions in nephelines are well documented by CL. Dull blue luminescence of nepheline cores may be caused by defects in the nepheline structure. The Fe content of the nepheline may also control the luminescence. Cores of nepheline contain less than 1% of the quenching element Fe, whereas an increasing Fe content in the rim (about 1.5 wt% FeO) points to increased quenching of the luminescence.

Most cores of apatites from the combeite-nephelinite studied exhibit the green luminescence normally associated with apatites from alkaline silicate rocks. Zoning in apatite reflects varying substitution of the REE and Mn^{2+} activators. However, a common feature of apatites from combeite-nephelinite at Oldoinyo Lengai are broad rims that luminesce with blue-violet colours, which is typical for carbonatitic environments (Mariano 1978; Koberski 1992).

Acknowledgements. A sample from the 1960 activity was kindly donated by Mike Le Bas. Discussions with Tony Mariano about carbonatites in general, and about cathodoluminescence in particular, were always enlightening. Funding for this research came from the Deutsche Forschungsgemeinschaft. Substantial help was also obtained from Freiburg University and from Wissenschaftliche Gesellschaft Freiburg in building up our cathodoluminescence equipment. The chapter has greatly benefited from thorough and constructive reviews by A.N. Mariano, P.L. Roeder, and K. Bell.

References

Carpenter AB, Oglesby TW (1976) A model for the formation of luminescently zoned calcite cements and its implications. Geological Society of America, Boulder, Abstr with Progr 8:469–470

Dawson JB (1962) Sodium carbonate lavas from Oldoinyo Lengai, Tanganyika. Nature 195: 1075–1076

Dawson JB (1989) Sodium carbonate lavas from Oldoinyo Lengai, Tanzania: Implications for carbonatite complex genesis. In: Bell K (ed) Carbonatites – genesis and evolution. Unwin Hyman, London, pp 255–277

Dawson JB, Garson MS, Roberts B (1987) Altered former alkalic carbonatite lava from Oldoinyo Lengai, Tanzania: inferences for calcite carbonatite lavas. Geology 15:765–768

Dawson JB, Pinkerton H, Norton G, Pyle D (1990) Physicochemical properties of alkali carbonatite lavas: data from the 1988 eruption of Oldoinyo Lengai. Geology 18:260–263

Dickson JAD (1980) Artificial colouration of fluorite by electron bombardment. Mineral Mag 43:820–822

Donaldson CH, Dawson JB, Kanaris-Sotiriou R, Batchelor RA, Walsh JN (1987) The silicate lavas of Oldoinyo Lengai, Tanzania. Neues Jahrb Miner Abh 156:247–279

Gittins J, McKie D (1980) Alkalic carbonatite magmas: Oldoinyo Lengai and its wider applicability. Lithos 13:213–215

Keller J (1989) Extrusive carbonatites and their significance. In: Bell K (ed) Carbonatites – genesis and evolution. Unwin Hyman, London, pp 70–88

Keller J, Krafft M (1989) Composition of natrocarbonatite lavas, Oldoinyo Lengai 1988. TERRA Abstr 1:286

Keller J, Krafft M (1990) Effusive natrocarbonatite activity of Oldoinyo Lengai, June 1988. Bull Volcanol 52:629–645

Koberski U (1992) Anwendungen der Kathodolumineszenz auf Fragestellungen in der Petrologie. Dr Thesis, University of Freiburg, 213pp (unpublished)

Marfunin AS (1979) Spectroscopy, luminescence, and radiation centers in minerals. Translated from the Russian publication by V.V. Schiffer. Springer, Berlin Heidelberg New York

Mariano AN (1978) The application of cathodoluminescence for carbonatite exploration and characterization. In: Proc 1st Int Symp on Carbonatites, Pocos de Caldas, Minas Gerais, Brasil. June 1976. Brasil Departemento Nacional da Producão Mineral, Brasilia, pp 39–57

Mariano AN (1988) Some further geological applications of cathodoluminescence. In: Marshall DJ (ed) Cathodoluminescence of geological materials. Unwin Hyman, London, pp 94–123

Mariano AN (1989) Cathodoluminescence emission spectra of rare earth element activators in minerals. In: Lipin BR, McKay GA (eds) Geochemistry and mineralogy of rare earth elements. Reviews in Mineralogy, vol 21. Mineral Soc Am, Washington DC, pp 339–348

Mariano AN, Roeder PL (1983) Kerimasi: a neglected carbonatite volcano. J Geol 91:449–455

Marshall DJ (1988) Cathodoluminescence of geological materials. Unwin Hyman, Boston, 146pp

Mason RA, Mariano AN (1990) Cathodoluminescence activation in manganese-bearing and rare earth-bearing synthetic calcites. Chem Geol 88:191–206

Meyers WJ (1974) Carbonate cement stratigraphy of the Lake Valley formation (Mississippian) Sacramento Mountains, New Mexico. J Sediment Petrol 44:837–861

Nakagawa M, Fukunaga K, Okada M, Atobe K (1988) Lattice defects in thermoluminescent calcite. J Lumin 40/41:345–346

Peterson TD (1989) Peralkaline nephelinites. I. Comparative petrology of Shombole and Oldoinyo Lengai, East Africa. Contrib Mineral Petrol 101:458–478

Peterson TD (1990) Petrology and genesis of natrocarbonatite. Contrib Mineral Petrol 105: 143–155

Sippel RF, Glover ED (1965) Structures in carbonate rocks made visible by luminescence petrography. Science 150:1283–1287

Teegarden K (1966) Halide lattices. In: Goldberg P (ed) Luminescence of inorganic solids. Academic Press, New York, pp 53–118

Nd and Sr Isotope Systematics
of the Active Carbonatite Volcano, Oldoinyo Lengai

K. Bell[1] and J.B. Dawson[2]

Abstract

The Nd and Sr isotopic compositions of natrocarbonatite, nephelinite, phonolite and plutonic blocks from Oldoinyo Lengai form a distinct negative correlation, and closely follow the East African Carbonatite Line (EACL). The $^{143}Nd/^{144}Nd$ ratios of the lavas range from 0.51249 to 0.51269, those of plutonic blocks from 0.51174 to 0.51270. The $^{87}Sr/^{86}Sr$ of the lavas range from 0.70414 to 0.70512; plutonic blocks range from 0.70378 to 0.70861. The nephelinites fall into two distinct groups; those with the more depleted isotopic signature include the combeite-bearing nephelinites. Published data from mantle xenoliths, both metasomatized and non-metasomatized, fall along the same array and suggest the interaction of two components, one enriched and the other slightly depleted relative to bulk Earth and CHUR. Isotopic variations indicate that the rocks from Oldoinyo Lengai cannot be derived by magmatic differentiation alone; at least two spatially related components are involved in the origin of rocks from Oldoinyo Lengai.

1 Introduction

The active carbonatite volcano, set in the Gregory Rift Valley of northern Tanzania, holds a unique place in igneous petrogenesis. In 1960, it erupted lava flows of sodium-rich carbonatite (natrocarbonatite) and this was the first documentation that carbonatites can be the products of active volcanicity (Dawson 1962). The significance of these lavas remains one of the more controversial issues in carbonatite research. Dawson (1966, 1989) and Le Bas (1981) consider them to be unmodified carbonate liquids, possibly derived by immiscibility from a parental carbonate-rich silicate parent; by alkali loss, these magmas could give rise to a wide range of carbonatite compositions. Conversely, Gittins (1989) suggests that natrocarbonatites are produced by fractionation of olivine sovite magma. Another point of dispute hinges on the uniqueness or otherwise of the Oldoinyo Lengai natrocarbonatite liquids, both in the context of whether similar lavas may have been erupted at other carbonatite volcanoes and whether, in plutonic carbonatite complexes, alkali metasomatism (fenitization) alongside

[1] Ottawa-Carleton Geoscience Centre, Department of Earth Sciences, Carleton University, Ottawa, Canada K1S 5B6
[2] Grant Institute of Geology, University of Edinburgh, West Mains Road, Edinburgh EH9 3JW, UK

calcite carbonatite intrusives is evidence for the presence of alkalies in the carbonatites at the time of their emplacement. These problems have been reviewed by Dawson et al. (1985), Dawson (1989) and Keller (1989).

The volcano is perhaps best known for its extrusions of alkali carbonatite lava (reviewed by Dawson et al., Chap. 1, this Vol.) and ash (Dawson et al. 1992), but the bulk of the cone is formed of two main units of nephelinitic and phonolitic tuffs and agglomerates (Dawson 1962, 1989): Unit 1, made up of older yellow tuffs and agglomerates (YTA) dated at 15 to 400 ka by correlation with the Ndutu and Naisiusiu Beds of the Olduvai Gorge succession (Hay 1989), and Unit 3, comprising younger black nephelinitic tuffs and agglomerates (BTA) that have been dated at 1250 to 2000 a on the basis of their correlation with the Namorod Ash of the Olduvai succession. Both units contain blocks of plutonic igneous rocks (jacupirangites, ijolites and nepheline syenites) presumed to derive from the subvolcanic complex beneath Oldoinyo Lengai, and also blocks of fenite and olivine-mica pyroxenites, respectively believed to result from metaso-matic alteration of crustal and mantle protoliths. The most recent activity at this volcano is of extrusions of alkali carbonatite lava and ash (Dawson et al., Chap. 1, this Vol.); however, there is evidence for carbonatite lava in the BTA (Dawson 1993) and distinctive white horizons in the YTA (Dawson 1962) may represent older carbonate ash deposits. Hence, although the modern activity has highlighted the latest extrusions of alkali carbonatite, it appears that it has been erupted sporadically at least during the later part of the eruptive history of the volcano.

As an individual volcano itself, however, Oldoinyo Lengai poses some intri-guing questions, some of which have recently been reviewed by Dawson (1989). First, are the natrocarbonatites pristine, unmodified carbonate liquids, or are they fractions of a more basic carbonate magma? Second, what is the relationship of the natrocarbonatites to the silicate magmatic rocks? Third, what is the relationship of the nephelinites to the phonolites? The most recent study of the silicate lavas (Donaldson et al. 1987) concluded that they could not be related by fractionation of any combination of the phenocryst phases, and that some other process, such as palagonitization or selective interaction with wall-rock, must have been involved. Fourth, are the natrocarbonatites entirely magmatic, or can we detect any involvement of either assimilated sedimentary trona (Milton 1968) or soda-rich brines (Eugster 1970)? The potential for answering some of these problems lies in a study of the Nd and Sr isotopic compositions of the lavas and the ejected plutonic blocks, since variations of Nd and Sr isotope ratios in combination with trace element geochemistry can provide certain constraints on magma sources and processes taking place during magma evolution.

For comparison, there already exists a large amount of Nd, Sr and Pb isotopic data from carbonatites from several different continents (e.g. Bell et al. 1987; Nelson et al. 1988; Bell and Blenkinsop 1989; Kwon et al. 1989) and these show a close similarity to data from oceanic islands, suggesting that the parental magmas to carbonatites are of mantle origin. Whether the parent melts are generated from the asthenosphere or ancient, depleted lithosphere is a subject of much debate and has been reviewed by Bell and Blenkinsop (1989). Few complexes have been studied in detail, but some, such as Oka (Wen et al. 1987), show very uniform Nd and Sr isotopic ratios, whereas others, such as Jacupiranga (Roden et al. 1985), Shombole (Bell and Peterson 1991) and Napak (Simonetti and Bell 1993) have significant internal isotopic variations and hence a complex history. Unravelling the evolutionary history of a carbonatite complex is difficult because of recrystallization of carbonate minerals, hydrothermal activity, repeated intrusions of carbonatite melts, metasomatism and late-stage mineralization.

2 Samples

The samples comprise both volcanic and plutonic magmatic rocks, and also two metasomatized rocks. The plutonic rocks occur as blocks in agglomerates, and indicate the presence of an igneous-metasomatic complex beneath the volcano. The samples are from most of the main volcano-stratigraphic units that form the volcano (Table 1) and petrographic details are given in the Appendix. The volcanic rocks are highly evolved olivine-free nephelinites and phonolites, and two samples of natrocarbonatite. The silicate lavas are representative samples of a gradational series, the detailed petrography, mineral chemistry and bulk chemistry of which are given by Donaldson et al. (1987); broadly, they are porphyritic and consist of nepheline, aegerine-augite, Ti-andradite and, in the

Table 1. Stratigraphy of Oldoinyo Lengai and sample localities

Unit 1	Yellow palagonitized nephelinitic and phonolitic tuffs and agglomerates, with rare interbedded flows of phonolite and nephelinite. This *oldest* unit comprises >90% of the total volume of the volcano Lava flows: BD 64[a] – nephelinite; BD 121[a] – phonolite Lava blocks: BD 29, BD 74, BD 91 – phonolites Xenoliths: BD 4A – jacupirangite; BD 77 – mica pyroxenite: 93 – olivine mica pyroxenite; BD 872 – eucolite-nepheline syenite[b]
Unit 2	Parasitic mica-rich tuff rings on the lower eastern and western slopes
Unit 3	Black nephelinitic tuffs and agglomerates Lava blocks: BD 50[a], BD 67[a] – phonolites; BD 54[a], BD 81[a], BD 126[a] – nephelinites Xenoliths: BD 33, BD 35, BD 45, BD 52, BD 122 – ijolites; BD 47 – wollastonite-rich ijolite
Unit 4	Minor flows of nephelinite in the summit area and from parasitic cones on the lower northern slopes Lava flows: BD 66, BD 119[a], BD 120[a] – nephelinites
Unit 5	Variegated carbonate-rich ashes of the summit area, northern and western slopes Xenolith: BD 343 – ijolite
Unit 6	Modern (post-1950) natrocarbonatite lavas and carbonate-silicate ashes Lava flows: BD 114 and BD 118[c]. 1960 eruption

[a] Petrography and bulk rock chemistry, and mineral chemistry data for some specimens, in Donaldson et al. (1987).
[b] Petrography and bulk rock chemical data in Dawson and Frisch (1971).
[c] Bulk rock data in Dawson (1962, 1989).

phonolites, sanidine phenocrysts, set in a matrix of nepheline and aegerine microphenocrysts, apatite, pyrrhotite, wollastonite, glass (sometimes palagonitized), and perovskite or titanite. The natrocarbonatite lavas, reviewed by Dawson (1989), consist of phenocrysts of nyerereite and gregoryite (complex Na, K, Ca carbonates containing considerable BaO, SrO, Cl, F, SO_3 and P_2O_5 in solid solution) in a matrix of apatite, gregoryite and nyerereite microphenocrysts, a (Mn, Fe) sulphide, sylvite, fluorite and barite. The magmatic plutonic rocks, ranging from jacupirangite to ijolite, show cumulate textures; they comprise varying proportions of clinopyroxene, nepheline, titano-magnetite, apatite, Ti-andradite, perovskite and, in some cases, wollastonite (Dawson 1989). Another group of rocks, represented by samples BD77 and BD93, are olivine mica pyroxenites that show replacement of olivine by pyroxene and phlogopite, and high concentrations of Mg, Ni and Cr. These samples are interpreted as metasomatized upper-mantle peridotites. A feature of some of these pyroxenites, but not in the two samples used in the present study, is that some blocks form cores to nephelinite or ijolite bombs; there has been interaction between the magma and the mica pyroxenite, resulting in mica elimination in the peripheral part of the block, together with K loss to the magma (Dawson and Smith 1994).

3 Analytical Results

Sr and Nd isotopic ratio measurements were made on a multicollector Finnigan MAT 261 solid source mass spectrometer at Carleton University. Samples were run as the chloride using a double filament technique for Nd and a single filament for Sr. Analyses are considered good to ±0.005% of the quoted values at the two sigma level. For full analytical details see Bell and Blenkinsop (1987a). Single analysis of the Eimer and Amend Sr standard measured during the study gave a value for the $^{87}Sr/^{86}Sr$ of 0.70802 ± 0.00003, and the La Jolla Nd standard a $^{143}Nd/^{144}Nd$ value of 0.51186 ± 0.00002. Uncertainties are given at the two-sigma level. Blanks during this study averaged 2 ng for Sr and 0.02 ng for Nd.

The analytical data are given in Table 2, and all are plotted in Fig. 1. Figure 1 also includes data from Bell and Peterson (1991) for four other Oldoinyo Lengai nephelinites collected on the northern slopes of the volcano by H. Eugster; the precise stratigraphic position of these four samples is not known. Modes, and mineral and bulk-rock chemistry for these samples, two of which (HOL-6 and -10) contain the rare NaCa silicate mineral, combeite, are given by Peterson (1989). Data from two natrocarbonatite flows and one combeite-bearing nephelinite sample from the June 1988 eruption are taken from Keller and Krafft (1990), and two natrocarbonatites (BD 114 and BD 118) from Bell and Blenkinsop (1987b).

Several features emerge from Fig. 1:

1. The isotopic data show a variation that is outside the limits of analytical uncertainty. Even excluding samples BD 93 and BD 77 (mica pyroxenites – the result of metasomatism of peridotite), the $^{87}Sr/^{86}Sr$ ratios vary from 0.70378 to

Table 2. $^{87}Sr/^{86}Sr$ and $^{143}Nd/^{144}Nd$ isotope ratios from Oldoinyo Lengai

Sample no.	Rock type	$^{87}Sr/^{86}Sr$	$^{143}Nd/^{144}Nd$	Reference
BD 114	c	0.70442	0.51261	Bell and Blenkinsop (1987b)
BD 118	c	0.70445	0.51259	"
OL 102	c	0.70437	0.51262	Keller and Krafft (1990)
OL 105	c	0.70439	0.51263	"
BD 54	n	0.70512	0.51249	This chapter
BD 64	n	0.70507	0.51249	"
BD 66	n	0.70418	0.51261	"
BD 81	n	0.70510	0.51249	"
BD 119	n	0.70435	0.51262	"
BD 120	n	0.70428	0.51263	"
BD 126	n	0.70495	0.51249	"
HOL-6	n	0.70429	0.51266	Bell and Peterson (1991)
HOL-10	n	0.70429	0.51266	"
HOL-14	n	0.70414	0.51269	"
HOL-16	n	0.70414	0.51269	"
OL 7	n	0.70434	0.51264	Keller and Krafft (1990)
BD 29	p	0.70434	0.51257	This chapter
BD 50	p	0.70460	0.51255	"
BD 67	p	0.70458	0.51250	"
BD 74	p	0.70462	0.51254	"
BD 91	p	0.70418	0.51264	"
BD 121	p	0.70435	0.51266	"
BD 4	j	0.70378	0.51270	"
BD 33	i	0.70439	0.51263	"
BD 35	i	0.70412	0.51268	"
BD 45	i	0.70450	0.51263	"
BD 47	i	0.70522	0.51249	"
BD 52	i	0.70425	0.51262	"
BD 122	i	0.70421	0.51266	"
BD 343	i	0.70441	0.51262	"
BD 77	px	0.70861	0.51174	"
BD 93	px	0.70529	0.51244	"
BD 872	ns	0.70418	0.51262	"

Rock types: c – carbonatite, n – nephelinite, p – phonolite.
Plutonic rocks: i – ijolite, j – jacupirangite, ns – nepheline syenite, px – pyroxenite.
$^{87}Sr/^{86}Sr$ ratios normalized to a $^{86}Sr/^{88}Sr$ ratio of 0.1194.
$^{143}Nd/^{144}Nd$ ratios normalized to a $^{146}Nd/^{144}Nd$ of 0.7219.

0.70522, whilst the $^{143}Nd/^{144}Nd$ ratios range from 0.51249 to 0.51270. Using the epsilon notation (De Paolo and Wasserburg 1976), Sr values range from -10.2 to $+10.2$, and Nd from $+1.2$ to -2.9. This range is one of the largest recorded for a single eruptive centre.

2. On a plot of $^{143}Nd/^{144}Nd$ vs $^{87}Sr/^{86}Sr$, the data lie along a linear array that extends from just inside the depleted quadrant to well within the enriched quadrant. The array is limited in the depleted quadrant by a cumulate jacupiran-gite and in the enriched quadrant by the two metasomatic pyroxenites. Much of the data cluster close to the values for a primitive undifferentiated reservoir,

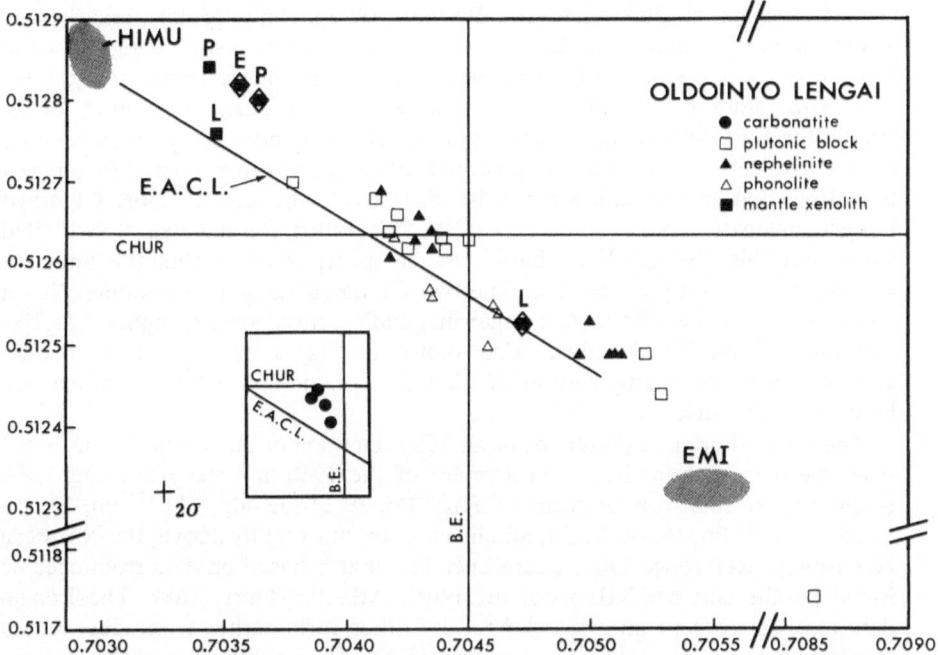

Fig. 1. Oldoinyo Lengai data. Symbols are indicated on diagram. *Straight line labelled E.A.C.L.* is the East African Carbonatite Line (Bell and Blenkinsop 1987b). *Inset* shows data from natrocarbonatites (Bell and Blenkinsop 1987b; Keller and Krafft 1990). Data from Cohen et al. 1984 for some mantle xenoliths (*filled squares*) or pyroxenes (*filled squares enclosed by diamonds*) separated from xenoliths. Spinel lherzolite xenoliths are from Eledoi (E), Pello (P) and Lashaine (L), northern Tanzania. Note the break in scale between $^{143}Nd/^{144}Nd = 0.5118$ and 0.5124, and between $^{87}Sr/^{86}Sr = 0.7055$ and 0.7085. (Approximate values for HIMU and EMI taken from Hart 1988)

defined in this work by a $^{143}Nd/^{144}Nd$ ratio for CHUR of 0.51264 and a bulk Earth $^{87}Sr/^{86}Sr$ ratio of 0.70450.

3. The nephelinites fall into two groups. The more depleted group (Group I) overlaps a set of magmatic plutonic rocks, mainly ijolites, of similar isotopic characteristics; all three combeite nephelinite samples belong to this more depleted group. The second, more enriched group (Group II), has ratios similar to those from ijolite, BD47. In short, there appear to be two isotopically distinct groups comprising both volcanic nephelinites and plutonic ijolites.

4. Neither of the two nephelinite groups are specific to any of the particular stratigraphic units shown in Table 1, and the range in isotopic values can be quite significant for lavas from any individual unit.

5. The Nd and Sr isotopic compositions of natrocarbonatite samples from the 1960 eruption, samples BD 114 and BD 118, are virtually identical to values quoted by Keller and Krafft (1990) for a natrocarbonatite from the June 1988 eruption.

6. Data from the phonolites straddle the two nephelinite groups, and there is a slight indication that the data fall into two sets, with one set having the same Sr isotopic ratio as some of the least depleted Group I nephelinites and ijolites.

7. The Oldoinyo Lengai array straddles and parallels, over much of its length, the East African Carbonatite Line (EACL), a linear array which is based on the Nd and Sr isotopic compositions of several young East African carbonatites (Bell and Blenkinsop 1987b). Relative to the EACL, most Oldoinyo Lengai magmatic rocks have higher $^{87}Sr/^{86}Sr$ ratios for a given $^{143}Nd/^{144}Nd$ value, but this divergence probably can be incorporated within the limits of analytical uncertainty of the line. The two Oldoinyo Lengai metasomites lie on an extension of the Oldoinyo Lengai magmatic array, toward higher $^{87}Sr/^{86}Sr$ and lower $^{143}Nd/^{144}Nd$ ratios. Also shown in Fig. 1 are data from mantle xenoliths from the nearby centres of Eledoi, Lashaine and Pello Hill which also lie close to the line.

The Pb isotope data (Williams et al. 1986; Dawson et al., Chap. 1, this Vol.) from the natrocarbonatite, from samples of the 1960 and the November 1988 eruptions, are relatively uniform ($^{208}Pb/^{204}Pb$, 39.14–39.40; $^{207}Pb/^{204}Pb$, 15.55–15.68; $^{206}Pb/^{204}Pb$, 19.19–19.26) and lie close to, but slightly above, the Northern Hemisphere Reference Line, a reference line that is based on data from oceanic island basalts and MORBs from the north Atlantic (Hart 1988). The Lengai data also lie close to an array defined by other carbonatites from East Africa (Grünenfelder et al. 1986) but are among the least radiogenic of any of the carbonatites analyzed. Grünenfelder et al. (1986) note that the slope of 0.098 ± 0.001 defined by the East African carbonatites is almost the same as the oceanic regression line of 0.102 (Tatsumoto 1978), and the data from most carbonatites extend along the complete array defined by OIBs. Such an array was interpreted by Grünenfelder et al. (1986) to be the result of mixing between two mantle components, one a LIL-element-depleted source, and the other a metasomatic fluid with a high $^{206}Pb/^{204}Pb$ ratio. The parent magmas to the East African carbonatites were derived from the same mantle source regions as OIBs, or from sources that have undergone very similar differentiation histories.

4 Discussion

The isotopic data (Fig. 1), clearly show that the Oldoinyo Lengai lavas, both silicate and carbonate, cannot be derived from a single, isotopically homogeneous parent. The simplest interpretation of the linear array of data is binary mixing of two mantle end members with similar Sr/Nd abundance ratios. Bell and Blenkinsop (1987b) considered the two end-members for the EACL to be either mineral phases or large-scale mantle reservoirs.

An alternative model is one that involves both assimilation and fractional crystallization, the so-called AFC model of DePaolo (1981), that is probably more realistic in terms of magma evolution than simple binary mixing. The AFC model requires selection of the degree of assimilation (r), which is the rate of mass assimilation to the rate of fractional assimilation and the solid-liquid partition coefficients (D) for the minerals that precipitate out of the melt. Roden et al. (1985) were able to model the distribution of data from the Jacupiranga complex in Brazil using the AFC model. We have made several attempts to use the AFC model to define a trend similar to the one shown in Fig. 1. Three parent magmas were chosen and five possible wall rocks, using two r values and two D values. The r values used were 0.2 and 0.4, and the D values were $D(Nd)$ = 0.7 and $D(Sr)$ = 0.4, and $D(Nd)$ = 0.5 and $D(Sr)$ = 0.5. F values (mass of magma body/initial mass of the magma) from 1 to 0.01 were generated. Two initial magmas were chosen to more or less parallel the treatment by Roden et al. (1985), a nephelinite from Hawaii ($^{143}Nd/^{144}Nd$ = 0.51303; $^{87}Sr/^{86}Sr$ = 0.70341) and a basalt from the Walvis Ridge ($^{143}Nd/^{144}Nd$ = 0.51269; $^{87}Sr/^{86}Sr$ = 0.70420). It was assumed that both contain the same Nd and Sr abundances of 54 ppm and 1140 ppm respectively. The third assumed magma was similar to HOL 14, a nephelinite from Oldoinyo Lengai. Contaminants included an average granulite, pelagic sediments, Archean granitoid rocks from the Tanzanian Shield, and average I and S type granites. A trend between the Hawaiian nephelinite and the granulite [r = 0.4, $D(Sr)$ = 0.5, $D(Nd)$ = 0.5] best fitted the main cluster of points from Oldoinyo Lengai for F values between 0.2 and 0.4. Variation of r from 0.2 to 0.4. generally has a minimal effect on the location of the AFC curve, although the position of the F values moves along the curve, and the positions of the F values tend to telescope. A crude fit to the Lengai trend can also be modelled by using an extremely high value for r of 0.9 and the granulite with 150 ppm Sr and 20 ppm Nd. In spite of the fact that the AFC model, in most cases, depends on using a set of arbitrarily chosen parameters, in general terms the calculations tell us that contamination of a mafic melt by lower crust may be one way of generating some of the silicate rocks from Lengai, although such an end-member would have to have a chemical composition compatible with the extreme peralkalinity and the undersaturation that characterize the Lengai lavas.

There is firm evidence that the upper mantle beneath northern Tanzania varies widely in its modal mineralogy and isotopic composition (Cohen et al. 1984; Dawson and Smith 1988); for example $^{87}Sr/^{86}Sr$ ratios of bulk rock peridotites and their phases range from 0.7036 to 0.8360, though the latter figure is exceptional. We suggest that the two olivine-mica pyroxenites BD77 and BD93, for which there is petrographic and bulk rock chemical evidence that they are "hybrids" between peridotite and K-rich melt (Dawson and Smith 1994), could be derived from this heterogeneous mantle. Relationships similar to these suggested ones are seen in situ between mica-rich veins and peridotite wall-rock in peridotite xenoliths in the Pello tuff-ring and Eledoi crater, some 8 km east of Oldoinyo Lengai (Dawson and Smith 1988). These xenoliths are interpreted as samples of veined and metasomatized upper-mantle beneath this sector of the rift valley. One mica-rich vein, in specimen BD 3847 from Pello Hill, has low $^{87}Sr/^{86}Sr$ and high $^{143}Nd/^{144}Nd$ ratios; a clinopyroxene from the wall-rock peridotite in BD 3847, and diopside in BD 128 (from Eledoi) are derived from a slightly more enriched source than the vein from BD 3847 (Cohen et al. 1984).

Interaction between two such components, a young asthenospheric melt and an ancient, enriched lithospheric peridotite, could theoretically give rise to isotopic values extending along the mantle array on the EACL, including the interpolation of the line to the most extreme of our samples, BD 77. Other possible mixing components also include lower crust and a mantle melt with a depleted signature, or depleted sub-continental lithospheric mantle with asthenosphere, in which case the latter would have to have an enriched signature.

We have evidence of interaction between young melts and metasomatized upper mantle in the case of many of the olivine-mica pyroxenites at Oldoinyo Lengai that form the cores to the nephelinite or ijolite bombs; there has been strong thermal metamorphic and metasomatic interaction between the pyroxenite blocks and the former melts (Dawson and Smith 1994). This suggests that during ascent, the Oldoinyo Lengai silicate magma had the opportunity to interact with upper mantle wall rocks. Indeed, the highly evolved nature of the nephelinites and phonolites (Donaldson et al. 1987) suggests residence in magma chambers during which evolution could take place. Assuming that these magma-chamber wall-rocks had isotopic characteristics of enriched mantle, as represented by BD 77 and BD 93, then the Oldoinyo Lengai data can be interpreted as a broad mixing line. In the case of the two main nephelinite groups, the differences could be explained by differing residence times, the more enriched group resulting from a long period of interaction with enriched peridotite wall rock. Overall, the least enriched Oldoinyo Lengai rock (jacupirangite BD 4A) has a more enriched source signature than many asthenosphere melts (more so than Ascension, St. Helena, Easter, Marion and the New England seamounts), perhaps indicating some interaction with an enriched mantle component.

The Sr and Nd isotopic data from Oldoinyo Lengai mimic the LoNd array of Hart et al. (1986) which was attributed to the mixing of HIMU and EMI mantle components (see Fig. 1), and this is certainly consistent with the Pb isotope ratios from the natrocarbonatites (Williams et al. 1986; Dawson et al., Chap. 4, this Vol.) which lie between the HIMU and EMI components in Pb isotope ratio diagrams. The wide range of Pb isotopic ratios from other East African carbonatites (Tilton and Bell, unpubl. data) is also consistent with the involvement of HIMU and EMI. HIMU may be ancient altered, subducted oceanic crust (Chase 1981; White and Hofmann 1982), mantle that has lost Pb to the core (Allègre 1982), or metasomatized mantle (Zindler and Hart 1986; Nakamura and Tatsumoto 1988). EMI may be recycled continental crust or lithosphere (Hawkesworth et al. 1984; Nakamura and Tatsumoto 1988; Gerlach et al. 1988) or metasomatized mantle (Zindler and Hart 1986). Hart et al. (1986) originally advocated a subcontinental lithospheric home for the EMI and HIMU components, with HIMU representing only slightly modified primitive planetary material represented by CHUR and bulk Earth. The lack, however, of EMI and HIMU signatures in lithospheric xenoliths and basalts that have interacted with lithosphere, led Hart (1988) to speculate that the EMI and HIMU components are metasomatic in origin and stored at the mantle/core boundary. No matter which of these two, quite different, models proposed by Hart et al. (1986) and Hart (1988) is accepted for the Oldoinyo Lengai lavas, of significance is the fact that both end members are spatially related in such a way that they can interact with one another, producing mixing arrays in isotope ratio diagrams, such as the one shown in Fig. 1.

5 Conclusions

Variation in the Nd and Sr isotopic compositions of the Oldoinyo Lengai lavas are significant, and rule out either simple melting of a homogeneous mantle or differentiation of a single parent magma. Binary mixing between two end members with similar Sr/Nd ratios or assimilation coupled with fractional crystallization best explain the distribution of the data on the ACP. Possible end-member pairs include: (1) depleted asthenosphere and metasomatized mantle, (2) depleted lithosphere and undifferentiated asthenosphere and (3) lower crust and an asthenospheric or depleted lithospheric melt. Further important features brought out by the isotopic data are the two groupings of nephelinites, and the position of the natrocarbonatite data on the ACP. The latter do not fall on any of the extreme positions of the linear array defined by the Oldoinyo Lengai data but lie close to the Group I nephelinites, the group with the most depleted mantle signature. Comparison of the data shown in Fig. 1 with data from Shombole (Bell and Peterson 1991), a 2-Ma nephelinite-carbonatite volcano that lies about 50 km north of Oldoinyo Lengai in Kenya, shows that the nephelinite values from Shombole are very similar to the Group I nephelinites and, more importantly, that the entire range of Nd and Sr isotopic compositions shown by the Shombole nephelinites is spanned by the data from the Shombole carbonatites. Because of the variable Nd and Sr isotopic compositions of nephelinite from Oldoinyo Lengai we feel that the isotopic signature of the 1960 and 1988 natrocarbonatite is perhaps but one of many that could have existed for carbonatite magmas throughout the eruptive history of Lengai, particularly since at least four phases of carbonatite activity have been visualized for Oldoinyo Lengai (Dawson 1962).

Acknowledgements. We would like to thank J.W. Card for help with the analyses and drafting, and J.W. Card and A. Simonetti for their comments on an earlier version of this manuscript. M.F. Roden and G.R. Tilton are acknowledged for careful reviews that improved the manuscript. This work was partly supported by NSERC operating grant A7813 awarded to KB.

Appendix

Petrographic notes

Lavas

The silicate lavas are porphyritic; the dominant phenocryst phase is nepheline; the matrix (<0.1 mm) consists of glass, devitrified glass, microlites of pyroxene, sanidine (in some), wollastonite and apatite, with tiny grains of pervoskite and Ti-magnetite. (Further details in Donaldson et al. 1987). Modes given below. All samples are BD series.

Sample no.	29	49	50	54	64	66	67	74	81	91	119	120	121
Phonolite(p) or nephelinite(n)	p	n	p	n	n	n	p	p	n	p	n	n	p
Groundmass	57	61	64	68	70	60	71	57	62	60	75	45	73
Nepheline	15	29	21	24	25	18	17	24	22	29	17	34	12
Pyroxene	4	2	6	7	5	10	3	7	7	8	3	3	2
Sanidine	13	tr.	8	0	0	0	8	11	0	0	0	0	12
Wollastonite	0	0	0	0	0	1	0	0	0	0	2	1	0
Garnet	0	0	0	0	0	8	0	0	2	0	3	1	0
Apatite	1	tr.	1	1	0	0	1	1	1	0	0	0	1
Titanite	tr.	0	0	1	0	0	tr.	0	1	0	0	0	0
Magnetite	0	0	0	0	0	0	0	0	0	0	0	1	0
Vesicles + cavities + glass	10	8	0	0	0	3	0	0	3	3	0	15	0
Xenoliths	0	0	0	0	0	0	0	0	2	0	0	0	0

Plutonic blocks (grain size > 2 mm)

BD 4A   Jacupirangite. Ti-augite 40%, perovskite 20%, Ti-magnetite 25%, apatite 15%

BD 33   Ijolite. Zoned cpx ($Di_{35-70}Hd_{20-35}Aeg_{6-27}$) 45%, garnet 5%, nepheline ($Ne_{75}Ks_{22}Qz_3$) 45%, others (magnetite, apatite, perovskite) 5%

BD 35   Mica-ijolite. Mica content variable (up to 10%), nepheline 33%, titanite + garnet 10%, pyroxene 55%, magnetite 2%. Pyroxene/mica aggregates after replaced olivine

BD 52   "Hybrid" ijolite. Similar to BD 122 but lacking megacrystic mica. Nepheline ($Ne_{71-75}Ks_{20}Qz_{6-9}$) 32%, pyroxene 52%, garnet 12%, Ti-magnetite and perovskite 3%, glass 1%

BD 77   Nepheline-veined mica pyroxenite, considered metasomatised upper mantle. Pyroxene ($Di_{75}Hd_9Aeg_{12}$) 50%, Ti-phlogopite (Mg_{82}) 45%, Ti-magnetite 3%, nepheline 2%. Close to the nepheline vein the pyroxene changes composition to ($Di_{68}Hd_{18}Aeg_6$)

BD 93   Mica-olivine pyroxenite. Strained and kinked olivine (Fo_{73}) 5%, pyroxene ($Di_{75}Hd_{11}Aeg_{10}$) 42%, Ti-phlogopite (Mg_{80}) 50%, Ti-magnetite 3%
         Both BD 77 and BD 93 show replacement textures

BD 122   "Hybrid" ijolite. Megacrysts of olivine, phlogopite and diopside. Olivine surrounded by diopside-mica-magnetite reaction rim. Megacryst mica and diopside overgrown by Ti- and Fe-rich equivalents

BD 343   Ijolite. Nepheline ($Ne_{76}Ks_{21}Qz_3$) 52%, pyroxene 31%, Ti-magnetite 9%, apatite 13%, glass 3%

BD 872   Eucolite-bearing nepheline syenite. Aegerine augite 48%, nepheline 20%, K-feldspar 22%, eucolite 5%, apatite 4%, titanite 1%. Nepheline, K-feldspar and eucolite poikilitic to pyroxene and titanite

References

Allègre CJ (1982) Chemical geodynamics. Tectonics 81:109–132

Bell K, Blenkinsop J (1987a) Archean depleted mantle – evidence from Nd and Sr initial isotopic ratios of carbonatites. Geochim Cosmochim Acta 51:291–298

Bell K, Blenkinsop J (1987b) Nd and Sr isotopic compositions of East African carbonatites: implications for mantle heterogeneity. Geology 15:99–102

Bell K, Blenkinsop J (1989) Neodymium and strontium isotope geochemistry of carbonatites. In: Bell K (ed) Carbonatites – genesis and evolution. Unwin Hyman, London, pp 278–300

Bell K, Peterson T (1991) Nd and Sr isotope systematics of Shombole volcano, East Africa, and the links between nephelinites, phonolites, and carbonatites. Geology 19:582–585

Bell K, Blenkinsop J, Kwon ST, Tilton GR, Sage RP (1987) Age and radiogenic isotopic systematics of the Borden carbonatite complex, Ontario, Canada. Can J Earth Sci 24:24–30

Chase CG (1981) Oceanic island Pb: two-stage histories and mantle evolution. Earth Planet Sci Lett 52:277–284

Cohen RS, O'Nions RK, Dawson JB (1984) Isotope geochemistry of xenoliths from East Africa: implications for development of mantle reservoirs and their interaction. Earth Planet Sci Lett 68:209–220

Dawson JB (1962) The geology of Oldoinyo Lengai. Bull Volcanol 24:349–387

Dawson JB (1966) Oldoinyo Lengai – an active volcano with sodium carbonatite flows. In: Tuttle OF, Gittins J (eds) Carbonatites. Wiley, New York, pp 155–168

Dawson JB (1989) Sodium carbonatite extrusions from Oldoinyo Lengai, Tanzania: implications for carbonatite complex genesis. In: Bell K (ed) Carbonatites – genesis and evolution. Unwin Hyman, London, pp 255–277

Dawson JB (1993) A supposed sövite from Oldoinyo Lengai, Tanzania: result of extreme alteration of alkali carbonatite lava. Mineral Mag 57:93–101

Dawson JB, Frisch T (1971) Eucolite from Oldoinyo Legai. Lithos 4:297–303

Dawson JB, Smith JV (1988) Veined and metasomatised peridotite xenoliths from Pello Hill, Tanzania: evidence for anomalously light mantle beneath the Tanzania sector of the East African Rift Vally. Contrib Mineral Petrol 100:510–527

Dawson JB, Smith JV (1994) Potassium loss during metasomatic alteration of mica pyroxenite from Oldoinyo Lengai, northern Tanzania: contrasts with fenitization. Contrib Mineral Petrol (in press)

Dawson JB, Garson MS, Roberts B (1985) Altered former alkalic carbonatite lava from Oldoinyo Lengai, Tanzania: inference for calcite carbonatite lavas. Geology 15:765–768

Dawson JB, Smith JV, Steele IM (1992) 1966 ash eruption of the carbonatite volcano Oldoinyo Lengai: mineralogy of lapilli and mixing of silicate and carbonate magmas. Mineral Mag 56:1–16

DePaolo DJ (1981) Trace element and isotopic effects of combined wall-rock assimilation and fractional crystallization. Earth Planet Sci Lett 53:189–202

DePaolo DJ, Wasserburg GJ (1976) Nd isotopic variations and petrogenetic models. Geophys Res Lett 3:249–252

Donaldson CH, Dawson JB, Kanaris-Sotiriou R, Batchelor RA, Walsh NJ (1987) The silicate lavas of Oldoinyo Lengai, Tanzania. Neues Jahrb Miner Abh 156:247–279

Eugster HP (1970) Chemistry and origin of the brines of Lake Magadi, Kenya. Mineral Soc Am Spec Pap 3:215–235

Gerlach DC, Cliff RA, Davies GR, Norry M, Hodgson N (1988) Magma sources of the Cape Verdes archipelago: isotopic and trace element constraints. Geochim Cosmochim Acta 52:2979–2992

Gittins J (1989) The origin and evolution of carbonatite magmas. In: Bell K (ed) Carbonatites – genesis and evolution. Unwin Hyman, London, pp 580–600

Grünenfelder MH, Tilton GR, Bell K, Blenkinsop J (1986) Lead and strontium isotope relationships in the Oka carbonatite complex, Quebec. Geochim Cosmochim Acta 50:461–468

Hart SR (1988) Heterogeneous mantle domains: signatures, genesis and mixing chronologies. Earth Planet Sci Lett 90:273–296

Hart SR, Gerlach DC, White WM (1986) A possible new Sr-Nd-Pb mantle array and con
    sequences for mantle mixing. Geochim Cosmochim Acta 50:1551–1557
Hawkesworth CJ, Rogers NW, van Calsteren PWC, Menzies MA (1984) Mantle enrichmen
    processes. Nature 311:331–335
Hay RL (1989) Holocene carbonatite-nephelinite tephra deposits of Oldoinyo Lengai, Tanzania
    J Volcanol Geotherm Res 37:77–91
Keller J (1989) Extrusive carbonatites and their significance. In: Bell K (ed) Carbonatites ·
    genesis and evolution. Unwin Hyman, London, pp 70–88
Keller J, Krafft M (1990) Effusive carbonatites activity of Oldoinyo Lengai, June 1988. Bul
    Volcanol 52:629–645
Kwon S-T, Tilton GR, Grünenfelder MH (1989) Lead isotope relationships in carbonatites an
    alkalic complexes: an overview. In: Bell K (ed) Carbonatites – genesis and evolution
    Unwin Hyman, London, pp 360–387
Le Bas MJ (1981) Carbonatite magmas. Mineral Mag 44:133–140
Milton C (1968) The "Natro-Carbonatite Lava" of Oldoinyo Lengai, Tanzania. Geol Soc Ar
    Program with Abstr, 202
Nakamura N, Tatsumoto M (1988) Pb, Nd, and Sr isotopic evidence for a multi-componen
    source for rocks of Cook-Austral islands and heterogeneities of mantle plumes. Geochir
    Cosmochim Acta 52:2909–2924
Nelson DR, Chivas AR, Chappell BW, McCulloch MT (1988) Geochemical and isotopi
    systematics in carbonatites and implications for the evolution of ocean-island source:
    Geochim Cosmochim Acta 52:1–17
Peterson TD (1989) Peralkaline nephelinites. I. Comparative petrology of Shombole an
    Oldoinyo Lengai, East Africa, Contrib Mineral Petrol 101:458–478
Roden MF, Murthy VR, Gaspar JC (1985) Sr and Nd isotopic composition of the Jacupirang
    carbonatite. J Geol 93:212–220
Simonetti A, Bell K (1993) Isotopic disequilibrium in clinopyroxenes from nephelinitic lava:
    Napak volcano, eastern Uganda. Geology 21:243–246
Tatsumoto M (1978) Isotopic composition of lead in oceanic basalt and its implication to mantl
    evolution. Earth Planet Sci Lett 38:63–87
Wen J, Bell K, Blenkinsop J (1987) Nd and Sr isotope systematics of the Oka Comple>
    Quebec, and their bearing on the evolution of the sub-continental upper mantle. Contri
    Mineral Petrol 97:433–437
White WM, Hofmann AW (1982) Sr and Nd isotope geochemistry of oceanic basalts an
    mantle evolution. Nature 296:821–825
Williams RW, Gill JB, Bruland KW (1986) Ra-Th disequilibria systematics: timescale c
    carbonatite magma formation at Oldoinyo Lengai volcano, Tanzania. Geochim Cosmochir
    Acta 50:1249–1259
Zindler A, Hart SR (1986) Chemical geodynamics. Annu Rev Earth Planet Sci 14:493–571

Stable Isotope Characteristics
of Recent Natrocarbonatites from Oldoinyo Lengai

J. KELLER[1] and J. HOEFS[2]

Abstract

Carbon and oxygen isotopic compositions of nyerereite and gregoryite phenocrysts and whole-rock samples of natrocarbonatite lavas from the June 1988 eruption of Oldoinyo Lengai lie within restricted ranges of $\delta^{13}C$ -6.3 to -7.1 and 5.8 to 6.7 for $\delta^{18}O$. These $\delta^{18}O$ and $\delta^{13}C$ values from unaltered natrocarbonatites and their carbonate phenocrysts support the conclusion that the carbonatitic magma was derived from the mantle and that their isotope composition was not changed by secondary isotopic exchange. Exposure to alteration under atmospheric conditions, weathering and hydration of the alkali carbonates at the surface produce distinctly higher $\delta^{18}O$, and also heavier $\delta^{13}C$ values. A recent natrocarbonatite exposed to weathering for only several weeks shows $\delta^{18}O$ and $\delta^{13}C$ values of 17.4 and -3.3‰, respectively.

1 Introduction

Oldoinyo Lengai is the only active carbonatite volcano on Earth and the isotopic composition of its recent lavas provides important information about the origin of carbonatites and the isotopic composition of the mantle source and its volatiles. Since the unique natrocarbonattic composition of the effusive eruptions of 1960–66 was recognized (Dawson 1962) the volcano became dormant until 1983 (Nyamweru 1988; Keller and Krafft 1990; Dawson et al., Chap. 1, this Vol.) and no sampling of unaltered natrocarbonatite was possible.

Recent renewed activity at Oldoinyo Lengai is characterized by quiet effusion of natrocarbonatitic lavas and by the presence of active carbonatitic lava lakes on a high crater platform (Keller and Krafft 1989, 1990; Dawson et al. 1990). As part of a study of recent activity of Oldoinyo Lengai during June 1988, completely fresh samples from flows and lava lakes were collected immediately during eruption. Keller and Krafft (1990) give a report of this activity, and petrologically characterize the analyzed samples of recent natrocarbonatites

---
[1] Institut für Mineralogie, Petrologie und Geochemie, Universität Freiburg, Albertstr. 23b, 79104 Freiburg, Germany
[2] Geochemisches Institut der Universität, Goldschmidtstr. 1, 37077 Göttingen, Germany

2 Previous Stable Isotope Studies of Oldoinyo Lengai Natrocarbonatites

Deines (1989) has given an extensive review of the present understanding of the behaviour of stable isotopes in carbonatite genesis and evolution. These studies have documented that primary igneous carbonatites range between -4.0 and $-8.0‰$ in $\delta^{13}C$ values (PDB) and between 6 and 10‰ in $\delta^{18}O$ values (SMOW), the so-called carbonatite box in Fig. 1 (Taylor et al. 1967; Sheppard and Dawson 1973; Hoefs 1987). This field is considered to represent the primary isotopic composition of typical carbonatite and, hence, of its mantle source. ("Primary" in this context is taken to mean carbonatites unaffected by surficial secondary processes; it is not meant to imply an unmodified partial mantle melt.) Higher values, in particular increases in $\delta^{18}O$, are commonly interpreted as being due to alteration processes and secondary isotopic exchange. Isotope fractionation during cooling and degassing of isotopically light C/O volatiles can also change the primary isotopic composition (Deines 1989). Regional differences pointing to isotopically heterogeneous sources of carbonatites have been noted by Deines and Gold (1973) and Nelson et al. (1988).

Carbon and/or oxygen isotope ratios for samples from the 1960/61 natrocarbonatite lavas from Oldoinyo Lengai have been reported by Denaeyer (1970),

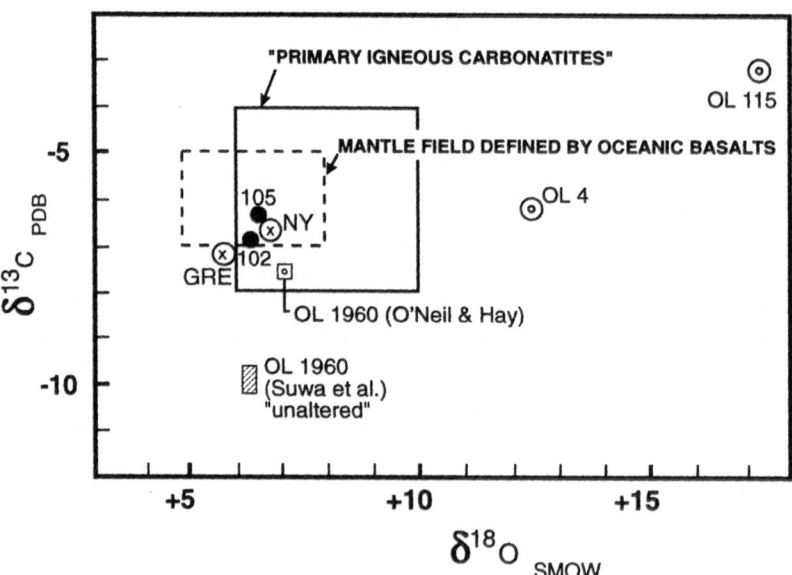

Fig. 1. Carbon and oxygen isotopic composition of fresh natrocarbonatite lava from Oldoinyo Lengai (OL 102 and OL 105), and of the separated phenocryst phases nyerereite NY and gregoryite GRE. Altered natrocarbonatites are OL 4 and OL 115. Comparison with published data from 1960/61 samples (O'Neil and Hay 1973; Suwa et al. 1975). Mantle fields based on "primary igneous carbonatites" (Taylor et al. 1967; Hoefs 1987), and primary C/O isotopes in basalts (Kyser 1986; Nelson et al. 1988; Deines 1989)

Vinogradov et al. (1970, 1971), Sheppard and Dawson (1973), O'Neil and Hay (1973), and particularly Suwa et al. (1975). The range of published values is −4.5 to −13‰ for $\delta^{13}C$ and 6 to 13‰ for $\delta^{18}O$. Such large variations for allegedly unaltered samples from the same eruption remained unexplained and, as a result, have hindered the interpretation of the data. The study of Suwa et al. (1975) of the lavas from the 1960/61 eruption is the most detailed, but the results obtained fell distinctly outside the expected field for primary carbonatites. In particular, the low $\delta^{13}C$ values posed problems in interpreting the stable isotopic data, since all of the delta values fell well below the field of primary igneous carbonatites (Fig. 1). A conclusion from the available data was that "the mean of $\delta^{13}C$ of the lavas from Oldoinyo Lengai differs from those of the rest of the carbonatites" (Deines 1989, p. 333). Suwa et al. (1975) attributed the low $\delta^{13}C$ values to partitioning of heavy carbon into a CO_2 gas phase during volcanic activity. On the other hand, the values of −7.6 for $\delta^{13}C$ and +7.1 for $\delta^{18}O$ for a lava sample of the same eruption, obtained by O'Neil and Hay (1973), fell in the field of primary carbonatites (Fig. 1) and suggested that the erratic isotope ratios of Suwa et al. (1975) were due to alteration of the analyzed natrocarbonatite lavas.

The first aim of the present study was, therefore, to analyze completely fresh natrocarbonatite samples from the 1988 effusive period, and to characterize the carbon and oxygen isotopic composition of the unaltered natrocarbonatites of Oldoinyo Lengai and changes during alteration.

Very low extrusion temperatures between 495 and 593 °C have been reported for the natrocarbonatites of Oldoinyo Lengai (Krafft and Keller 1989; Dawson et al. 1990). To understand possible fractionations of C and O isotopes during crystallization and solidification in this temperature range, the isotopic compositions of the carbonate phenocrysts, nyerereite and gregoryite, were determined and compared with whole-rock and matrix compositions.

During the collection of lavas in June 1988, gas samples were also taken directly from active natrocarbonatite vents. Javoy et al. (1989) reported the isotopic composition of CO_2 from these gases as −2.62 ± 0.02 for $\delta^{13}C$ and 12.9–18.3 for $\delta^{18}O$. Javoy et al. (1989) also published preliminary C and O isotope ratios for a lava sample obtained with the gas sampling device. These preliminary results were similar to the data of O'Neil and Hay (1973) and contrasted to those of Suwa et al. (1975).

A controversial hypothesis for the unusual natrocarbonatitic compositions of Oldoinyo Lengai was one involving trona sediments from nearby Lake Natron (Milton 1968; Eugster 1970). Bell and Dawson (Chap. 7, this Vol.) have given a thorough review of the trona discussion. On the basis of the stable isotopic composition of trona sediments presented by O'Neil and Hay (1973) and Suwa et al. (1975), this hypothesis is difficult to accept, despite the recent attestation by Milton (1989). Moreover, Nd- and Sr-isotopes of natrocarbonatites and associated silicate lavas are compelling evidence against any sediment involvement in their magma genesis (Bell and Blenkinsop 1989; Dawson 1989; Keller and Krafft 1990).

It is widely agreed that carbonatites are ultimately derived from the mantle (Wyllie et al. 1990). This includes the natrocarbonatites of Oldoinyo Lengai

(Peterson and Kjarsgaard, Chap. 11, this Vol.). Therefore, the isotopic composition of freshly erupted natrocarbonatite is an important key to the composition of the mantle source. If the assumption is made that natrocarbonatites and calciocarbonatites are both the result of different fractionation paths of a parent silicate melt, leading to varying degrees of peralkalinity (Kjarsgaard et al., Chap. 12, this Vol.), then the stable isotope composition of the completely fresh natrocarbonatites from Oldoinyo Lengai might be used to define the primary isotopic signature of carbonatites.

3 Petrographic and Chemical Studies of Analyzed Samples

Major and trace elements for fresh lava samples of Oldoinyo Lengai from June 1988 are given in Keller and Krafft (1990). Due to the deliquescent nature of natrocarbonatite, all samples collected in June 1988 were taken immediately during eruption and sealed before complete cooling. The compositions for all samples analyzed in this study for their oxygen and carbon isotopes are reported in Table 1.

Petrographic descriptions of Oldoinyo Lengai natrocarbonatites from the 1960/61 eruption are given by Dawson (1962, 1989), Cooper et al. (1975) and Peterson (1990). The studies of the 1988 lavas by Keller and Krafft (1990) and Dawson et al. (1990) have established that the compositions are similar to those from the earlier lava eruptions. The porphyritic lavas contain, on average, about 50% phenocrysts of the Na-Ca-carbonates, nyerereite and gregoryite. Tabular nyerereite and more oval-shaped gregoryite phenocrysts occur in about equal amounts and range in size between 1–2 mm (Koberski and Keller, Chap. 6, this Vol.). The crystallization sequence shows that both phases crystallized contemporaneously with gregoryite possibly starting slightly earlier. Both minerals are water-soluble and highly unstable under atmospheric conditions, but are perfectly fresh in the samples used in this study. Microprobe data for nyerereite and gregoryite are reported for the 1988 lavas by Keller and Krafft (1990) and for the lavas of the 1960–66 eruption by Peterson (1990). They show the average composition of $(Na_{0.82}K_{0.19})_2(Ca,Sr,Ba)_{0.98}(CO_3)_2$ for nyerereite and $Na_{1.74}K_{0.1}(Ca,Sr,Ba)_{0.16}CO_3$ for gregoryite. Complex substitution of CO_3 by SO_4, PO_4, F and Cl is important, particularly in gregoryite (Table 1). The groundmass of all lavas consists of microphenocrysts and microlites of nyerereite and gregoryite set in a matrix consisting of a fine-grained granular or sheaf-like intergrown domains of sylvite and fluorite (Keller and Krafft 1990; Koberski and Keller, Chap. 6, this Vol.). Additional groundmass phases are Fe-alabandite, apatite, MgF_2 (sellaite) and $BaCO_3$, and volumetrically very minor accessory oxides, halides, and carbonates of complex composition (Keller and Krafft 1990; Dawson et al., Chap. 4, this Vol.).

Samples OL 102 and OL 105 of this study (Table 1) are representatives of the main textural varieties of natrocarbonatite represented by highly porphyritic, and aphyric (residual liquid) lavas, respectively (Keller and Krafft 1990; Keller and Spettel, Chap. 5, this Vol.).

Table 1. Carbon and oxygen isotopic composition of fresh natrocarbonatite lava from Oldoinyo Lengai, from separated phenocryst phases, and from altered natrocarbonatites Chemical compositions characterize the analyzed samples

NO.	OL 102	OL 105	NY	GRE	OL 115	OL 4
Stable isotopic compositions						
$\delta^{13}C$ PDB	−6.87	−6.30	−6.65	−7.14	−3.29	−6.20
$\delta^{18}O$ SMOW	+6.32	+6.49	+6.74	+5.78	+17.38	+12.39
SiO_2	0.16	0.21	–	–	0.38	1.06
TiO_2	0.02	0.02	–	–	0.04	0.04
Al_2O_3	bd	bd	–	–	bd	0.05
Fe_2O_3	0.28	0.48	0.03	0.06	0.57	1.43
MnO	0.38	0.60	0.10	0.12	0.81	0.74
MgO	0.38	0.52	–	–	0.77	bd
CaO	14.02	12.86	23.84	7.48	23.33	27.19
SrO	1.42	1.43	2.25	0.61	3.07	1.28
BaO	1.66	2.17	0.81	0.27	3.57	1.11
Na_2O	32.22	30.42	23.56	45.67	20.80	21.91
K_2O	8.38	9.14	8.39	3.82	3.75	1.20
P_2O_5	0.85	0.75	0.56	0.96	0.92	0.99
CO_2	31.55	28.12	38.96	35.18	29.25	30.30
Cl	3.40	5.18	0.26	0.51	0.57	0.10
F	2.50	4.10	0.06	0.20	6.08	1.40
SO_3	3.72	5.58	1.15	4.70	1.52	1.03
H_2O^+	0.56	0.44	–	–	6.95	10.32
$-O = F,Cl$	−1.82	−2.90	−0.08	−0.20	−2.69	−0.61
Total	99.68	99.12	99.88	100.38	99.69	99.54
$^{87}Sr/^{86}Sr^a$		0.70437	0.70439			
$^{143}Nd/^{144}Nd$		0.51262	0.51263			

OL 102 Porphyritic lava, June 25, 1988 (Keller and Krafft 1990).
OL 105 Aphyric lava, June 25, 1988 (Keller and Krafft 1990).
NY   Nyerereite phenocrysts of OL 102.
GRE   Gregoryite phenocrysts of OL 102.
OL 115 Porphyritic natrocarbonatite lava, exposed to the atmosphere for an estimated period of some weeks.
OL 4   Pre-1960 nyerereite–carbonatite, hydrated to pirssonite–nyerereite carbonatite, location, Barry's Pinnacles.
[a] Sr and Nd isotopic data by K. Bell (in Keller and Krafft 1990).

The effects of secondary alteration and recrystallization on the stable isotope ratios of carbonatites are essential to the interpretation of the δ values different to those from primary carbonatites. The rapid hydration of the lavas of Oldoinyo Lengai provides an opportunity for testing empirically the isotopic changes that take place from the start of atmospheric alteration to the stage of complete isotopic exchange with meteoric waters. The major effects of alteration are dissolution of soluble minerals and hydration of nyerereite to pirssonite [$Na_2Ca(CO_3)_2 \cdot 2H_2O$] and gaylussite [$Na_2Ca(CO_3)_2 \cdot 5H_2O$]. Atmospheric humidity results very rapidly in white efflorescences of hydrated sodium carbonates and sylvite on the dark grey to black natrocarbonatite lavas. Within 1 or 2 days the flows look light grey and become covered with white incrustations. During the visit to Oldoinyo Lengai in June 1988, about one third of the crater floor became flooded with fresh lava flows. The remaining area of the summit crater was covered with older flows which were strongly altered but which perfectly preserved flow surfaces and flow outlines of mostly ropy pahoehoe-type and clinkery aa-type lavas. However, the material was brownish, friable, soft and earthy. It can be deduced from the continuous filling of the former explosion crater, and from the reports collected by Nyamweru (1988), that these flows were only a few weeks or months old (Keller and Krafft 1990).

One sample (OL 115) of altered natrocarbonatite, an earthy pseudomorph of pahoehoe crust, was analyzed in detail in order to understand the compositional and isotopic changes that occur during weathering. The texture can still be recognized as a porphyritic natrocarbonatite, identical to the texture of fresh material. Sylvite is completely dissolved, and most of the gregoryite has disappeared. Some nyerereite is still present. However, the dominant phase is secondary gaylussite, the hydrated form of nyerereite. Drying the rock powder at 105 °C converts all of the gaylussite to the less hydrated equivalent, pirssonite. The composition of this altered natrocarbonatite, OL 115, is reported in Table 1 and can be compared with that of fresh porphyritic lava, OL 102. The most obvious changes are the loss of Na, K, Cl and SO_3, combined with an increase in H_2O, CaO, SrO, BaO, F, and Mn. Similar changes are obvious in the nyerereite-pirssonite carbonatite, OL 4 (Table 1). This sample was collected from the only known pre-1960 natrocarbonatite at Lengai, possibly related to the crater platform which existed prior to the explosive eruption of 1917. This pirssonite-natrocarbonatite was first described and analyzed by Dawson et al. (1987). OL 4 is an example of an originally gregoryite-free natrocarbonatite that contained much less KCl, if any, than the recent lavas. Its alteration is much slower than that shown by the more recent lavas, and is dominated by partial hydration of nyerereite to pirssonite. The textural features of this hydration can be studied with cathodoluminescence and are shown in Fig. 1 of Koberski and Keller (Chap. 6, this Vol.).

4 Analytical Techniques

Isotopic compositions are reported for whole-rock natrocarbonatites. The porphyritic main facies is represented by sample OL 102, and the aphyric by sample OL 105. Phenocryst phases nyerereite and gregoryite have been separated mechanically and hand picked. The two altered natrocarbonatites, OL 115 and OL 4, are powdered bulk samples, dried at 105 °C without any further treatment. Analytical methods used for the chemical data in Table 1 are reported in Keller and Krafft (1990).

For the carbon and oxygen isotope determination all analyzed samples were treated with 100% phosphoric acid. The liberated CO_2 was measured mass spectrometrically with a Finnigan MAT 251. The results of the oxygen and carbon isotope analyses are given in the usual permil deviation relative to a standard. All $\delta^{18}O$ values are given relative to SMOW, $\delta^{13}C$ values relative to PDB. The reproducibility of both δ values is better than $\pm$ 0.2‰.

5 Results and Discussion

$\delta^{13}C$ and $\delta^{18}O$ data are shown in Table 1 and Fig. 1. The isotopic compositions of both types of whole-rock lavas, phenocryst-rich and aphyric varieties, respectively, are virtually identical. The phenocryst phases of natrocarbonatite, nyerereite and gregoryite, which had not been analyzed for their isotopic composition before, show very similar ^{13}C and ^{18}O contents to the bulk lavas. $\delta^{13}C$ values of the fresh lava and of both carbonate phenocryst phases vary within a narrow range of -6.3 to -7.1‰. $\delta^{18}O$ values of natrocarbonatite lavas and phenocrysts range from 5.8 to 6.7‰. Our data for the lavas of 1988 differ from the results for the 1960 lavas of Suwa et al. (1975), but are close to the values for the 1960–61 lavas given by O'Neil and Hay (1973). There is no explanation for the values obtained by Suwa et al., especially their low $\delta^{13}C$. The detailed analyses of lavas and phenocryst phases, together with additional stable isotope data in Dawson et al. (Chap. 1, this Vol.) and the C and O isotopic ratios on lavas from a 1985 effusion by Hay (1989), define the average primary oxygen and carbon isotopic composition of the natrocarbonatites of Oldoinyo Lengai at $+6.6$ for $\delta^{13}C$ and $+6.4$ for $\delta^{18}O$. As a result, the earlier data of Suwa et al. cannot be used for the discussion of primary natrocarbonatite composition.

All isotopic values presented are at the lower, primitive end of the carbonatite range. In particular, $\delta^{18}O$ values $< +6$ have rarely been reported for primary carbonatites. Such low values were obtained for early formed calcite megacrysts of porphyritic alvikites (Hubberten et al. 1988). Deines (1989) proposes a filter of $+5.5-8.5$‰ for carbonatitic $\delta^{18}O$ values not disturbed by secondary processes.

Deines and Gold (1973) have also observed a larger variation of $\delta^{13}C$ and $\delta^{18}O$ in shallow level or volcanic carbonatites compared to carbonatites from deep-seated intrusions. The constant and low carbon and oxygen δ values of the volcanic carbonatites of Oldoinyo Lengai are at variance with this observation. These results suggest that the higher oxygen isotope values outside the field of "primary igneous carbonatites", found in a number of carbonatite complexes, must be more generally interpreted as due to secondary effects such as recrystallization and re-equilibration after emplacement, in spite of the fact that protracted Rayleigh fractionation can produce a similar trend to that resulting from alteration of fresh natrocarbonatite (Deines 1989).

Degassing of ^{13}C-enriched CO_2 could account for magmatic melts having somewhat lighter $\delta^{13}C$ values than their source. However, the fractionation between $CaCO_3$ and CO_2 is rather small (Bottinga 1968) and mass balance considerations suggest that degassing of CO_2 should have a minor effect in changing the primary carbon isotope composition. For Oldoinyo Lengai the isotopic composition of the CO_2 gas phase is -2.62‰ $\delta^{13}C$, and $+12.9$ to 18.3‰ $\delta^{18}O$ (Javoy et al. 1989). This unusually large isotopic fractionation between natrocarbonatite lava and CO_2 gas phase and the constant composition of lavas and phenocrysts suggest that the amount of CO_2 exsolved from the carbonatite magma during its crystallization was not sufficient to change the isotopic composition of the reservoir.

The low $\delta^{18}O$ values for the natrocarbonatite samples, which are the only known carbonatites for which secondary alteration can be excluded, may further imply that the so-called primary carbonatite box of Taylor et al. (1967), or the range of +6 to +8‰ $\delta^{18}O$ vs. SMOW given by Deines and Gold (1973) for unaltered carbonatites can be narrowed down to values of 5.5–7‰ $\delta^{18}O$. This is confirmed by $\delta^{18}O$ values given by Kyser et al. (1982) for basaltic lavas, the large majority of which fall in the range between +5.0 and +6.2‰ vs. SMOW (Kyser et al. 1982; Nelson et al. 1988). The measured natrocarbonatite values are among the lowest $^{18}O/^{16}O$ reported for primary igneous carbonatites, but are still slightly heavier than silicate mantle minerals. Because carbonates are generally heavier in ^{18}O than the typical mantle minerals, olivine and pyroxene, with which they are in equilibrium (e.g. Kyser 1986; Hoefs 1987), such a $^{18}O/^{16}O$ enrichment is in line with theoretical expectations.

The reported $\delta^{13}C$ values are characteristic of many primary carbonatites and are supposed to represent the carbon isotopic composition of their mantle source. $\delta^{13}C$ values in the range of −5 to −6‰ characterize important mantle-carbon reservoirs, e.g. the one for most OIBs and mid-ocean ridge basalts (Des Marais and Moore 1984; Kyser 1986), diamonds and carbonates in kimberlites (Deines 1989; Javoy et al. 1986). Small-scale regional differences are observed and discussed in terms of heterogeneities in the source (Kyser 1986; Nelson et al. 1988; Deines 1989). Based on the data of Suwa et al. (1975), Oldoinyo Lengai was quoted as an important example of these regional variations.

Comparison of whole-rock natrocarbonatite isotope values with early formed phenocrysts of gregoryite and nyerereite (Table 1; Fig. 1) shows only very small differences. Isotopic fractionation between both carbonate phases, and between carbonate phenocrysts and residual melt is thus of very minor importance in the system under investigation. Gregoryite, the phenocryst phase that was the earliest mineral to crystallize, shows the lowest $\delta^{13}C$ and $\delta^{18}O$ and only a slight fractionation between gregoryite and nyerereite might be suggested from the data.

Another important aspect of our results is the distinctly higher $\delta^{18}O$ of +17.4 and +12.4 in samples OL 115 and OL 4. These samples were selected because of their obvious meteoric alteration. The rapid increase in $\delta^{18}O$ underlines low susceptible carbonatites, natrocarbonatites in particular, are to low temperature isotopic exchange. ^{13}C is less affected in the beginning alteration (OL 4), but looses its mantle signature with the stronger alteration effects in OL 115 leading to a $\delta^{13}C$ of −3.3. The different degree of isotopic re-equilibration of the two altered samples is also underlined by the higher ^{18}O content of OL 115 compared to OL 4. The data presented for these two samples, specifically selected for evaluating the effects of alteration, are complemented by a series of oxygen and carbon isotopic compositions given by Hay (1989) for altered natrocarbonatite clasts, and calcrete and calcite formed on a substrate of natrocarbonatitic tuffs from Lengai. The $\delta^{18}O$ values range from +25.8 to +33.7 and have almost completely exchanged their oxygen with meteoric water. The $\delta^{13}C$ values in altered natrocarbonatite lapilli are -3.5 and -4.1 (Hay 1989, Table 5) and indicate similar degree of carbon exchange as OL 115 ($\delta^{13}C$ $-3.3‰$, Table 1) during weathering.

6 Conclusions

Carbon and oxygen isotopic compositions of Oldoinyo Lengai natrocarbonatites have isotopic signatures of primary mantle-derived carbon and carbonates ($+6.58$ for $\delta^{13}C$ and $+6.40$ for $\delta^{18}O$). The results of Suwa et al. (1975) are not confirmed. Despite the low extrusion temperatures in natrocarbonatite volcanism, no significant isotope fractionation has been detected between melt and carbonate phenocryst phases. Thus, stable isotope data of very fresh lava samples are clear evidence for the unaltered mantle origin of the carbonate. They rule out any significant sediment involvement in natrocarbonatite genesis, but cannot independently discriminate between different models of natrocarbonatite petrogenesis such as primary partial melt products, fractionation or immiscible separation.

Natrocarbonatites from Oldoinyo Lengai narrow down the field of primary igneous carbonatites to isotopic values close to the low $\delta^{18}O$ field of different mantle reservoirs, such as those from which MORB and OIB basalts or diamonds are generated.

Exposure of natrocarbonatites to atmospheric alteration, and hydration of the alkalicarbonate phases readily changes the stable isotope composition. Oxygen, in particular, is changed to heavier $\delta^{18}O$-values, and it is concluded that low $\delta^{18}O$ carbonatites ($\delta^{18}O$ 5.5–8‰) are unlikely to have experienced low temperature alteration and therefore represent isotopically primary igneous carbonatites.

Acknowledgements. The stimulus provided by Katja and Maurice Krafft during the Oldoinyo Lengai expedition will always be treasured. Celia Nyamweru is thanked for her advice and help. The Peterson brothers of Dorobo Safaris at Arusha provided perfect logistical support. The paper has greatly benefited from suggestions and reviews by P. Deines, K. Kyser and Keith Bell.

References

Bell K, Blenkinsop J (1989) Neodymium and strontium isotope geochemistry of carbonatites. In: Bell K (ed) Carbonatites – genesis and evolution. Unwin Hyman London, pp 278–300

Bottinga Y (1968) Calculation of fractionation factors for carbon and oxygen isotopic exchange in the system calcite-carbon dioxide-water. J Phys Chem 72:800–808

Cooper AF, Gittins J, Tuttle OF (1975) The system Na_2CO_3-K_2CO_3-$CaCO_3$ at 1 kilobar and its significance in carbonatite petrogenesis. Am J Sci 275:534–560

Dawson JB (1962) Sodium carbonate lavas from Oldoinyo Lengai, Tanganyika. Nature 195: 1075–1076

Dawson JB (1989) Sodium carbonate lavas from Oldoinyo Lengai, Tanzania: implications for carbonatite complex genesis. In: Bell K (ed) Carbonatites – genesis and evolution. Unwin Hyman, London, pp 255–277

Dawson JB, Garson MS, Roberts B (1987) Altered former alkalic carbonatite lava from Oldoinyo Lengai, Tanzania: inferences for calcite carbonatite lavas. Geology 15:765–768

Dawson JB, Pinkerton H, Norton GE, Pyle DM (1990) Physicochemical properties of alkali carbonatite lavas: data from the 1988 eruption of Oldoinyo Lengai. Geology 18:260–263

Deines P (1989) Stable isotope variations in carbonatites. In: Bell K (ed) Carbonatites – genesis and evolution. Unwin Hyman, London, pp 301–359

Deines P, Gold DP (1973) The isotopic composition of carbonatite and kimberlite carbonates and their bearing on the isotopic composition of deep-seated carbon. Geochim Cosmochim Acta 37:1709–1733

Denaeyer ME (1970) Rapports isotopiques δO et δC et conditions d'affleurement des carbonatites de l'Afrique Central. C R Acad Sci 270D:2155–2158

Des Marais DJ, Moore JG (1984) Carbon and its isotopes in mid-oceanic basaltic glasses. Earth Planet Sci Lett 69:43–57

Eugster H (1970) Chemistry and origin of the brines of Lake Magadi, Kenya. Mineral Soc Am Spec Pap 3:215–236

Hay RL (1989) Holocene carbonatitie-nephelinite tephra deposits of Oldoinyo Lengai, Tanzania. J Volcanol Geotherm Res 37:77–91

Hoefs J (1987) Stable isotope geochemistry. 3rd edn. Springer, Berlin Heidelberg New York, 241 pp

Hubberten HW, Katz-Lehnert K, Keller J (1988) Carbon and oxygen isotope investigations in carbonatites and related rocks from the Kaiserstuhl, Germany. Chem Geol 70:257–274

Javoy M, Pineau F, Delorme H (1986) Carbon and nitrogen isotopes in the mantle. Chem Geol 57:41–62

Javoy M, Pineau F, Staudacher T, Cheminée JL, Krafft M (1989) Mantle volatiles sampled from a continental rift: the 1988 eruption of Oldoinyo Lengai. TERRA Abstr 1:324

Keller J, Krafft M (1989) Composition of natrocarbonatite lavas, Oldoinyo Lengai 1988. TERRA Abstr 1:286

Keller J, Krafft M (1990) Effusive natrocarbonatite activity of Oldoinyo Lengai, June 1988. Bull Volcanol 52:629–645

Krafft M, Keller J (1989) Temperature measurements in carbonatite lava-lakes and flows: Oldoinyo Lengai, Tanzania. Science 245:168–170

Kyser TK (1986) Stable isotope variations in the mantle. In: Valley JW et al. (eds) Stable Isotopes in high temperature geological processes. Rev Mineral 16:141–164

Kyser TK, O'Neil JR, Carmichael ISE (1982) Genetic relations among basic lavas and ultramafic nodules: evidence from oxygen isotopic compositions. Contrib Mineral Petrol 81:88–102

McKie D, Frankis EJ (1977) Nyerereite: a new volcanic carbonate mineral from Oldoinyo Lengai. Z Kristallogr 145:73–95

Milton C (1968) The "Natro-Carbonatite Lava" of Oldoinyo Lengai, Tanzania. Geol Soc Am Program with Abstr, 202

Milton C (1989) The Oldoinyo Lengai natrocarbonatite. Int Geol Congr Washington 1989, Abstr

Nelson DR, Chivas AR, Chappell BW, McCulloch MT (1988) Geochemical and isotopic systematics in carbonatites and implications for the evolution of ocean-island sources. Geochim Cosmochim Acta 52:1–17

Nyamweru C (1988) Activity of Oldoinyo Lengai volcano, Tanzania, 1983–1987. J Afr Earth Sci 7:603–610

O'Neil JR, Hay RL (1973) $^{18}O/^{16}O$ ratios in cherts associated with the saline lake deposits of East Africa. Earth Planet Sci Lett 19:257–266

Peterson TD (1990) Petrology and genesis of natrocarbonatite. Contrib Mineral Petrol 105:143–155

Sheppard SMF, Dawson JB (1973) $^{13}C/^{12}C$ and D/H isotope variations in "primary igneous carbonatites". Fortschr Mineral 50:128–129

Suwa K, Oana S, Wada H, Osaki S (1975) Isotope geochemistry and petrology of African carbonatites. Phys Chem Earth 9:735–745

Taylor HP Jr, Frechen J, Degens ET (1967) Oxygen and carbon isotope studies of carbonatites from the Laacher See district, West Germany and the Alnö district, Sweden. Geochim Cosmochim Acta 31:407–430

Vinogradov AP, Kropotova OI, Gerasimovsky VI (1970) Carbon isotope composition for carbonatites from East Africa. Geokhimiya 1970:643–646

Vinogradov AP, Dontsova EI, Gerasimovsky VI, Kuznetsova LD (1971) Oxygen isotopic composition of carbonatites in the rift zones of East Africa. Geochemistry Int 8:307–313

Wyllie PJ, Baker MB, White BS (1990) Experimental boundaries for the origin and evolution of carbonatites. Lithos 26:3–19

Decay Series Evidence for Transfer and Storage Times of Natrocarbonatite Magma

D.M. Pyle

Abstract

Short-lived radioactive disequilibria between nuclides of the ^{238}U and ^{232}Th decay series measured in the natrocarbonatite lavas of Oldoinyo Lengai provide unique constraints on the timing and mechanism of magmagenesis. Measured disequilibria are consistent with an origin of natrocarbonatite by immiscible exsolution of between 5 and 25 wt% carbonatite from a silicate melt. The particular parent silicate magma cannot be definitively constrained; it could be a peralkaline nephelinite, or intermediate between nephelinite and phonolite.

Disequilibria between ^{232}Th-^{228}Ra-^{228}Th constrain the time scale of magma generation, ascent and evolution. The available data are consistent with, but do not prove, the existence of a steady-state constant-volume magma reservoir below Oldoinyo Lengai. The 1960 and 1963 lava compositions indicate that this reservoir had a mean lifetime τ of 10–15 years. The 1960–1963 magma at the time of eruption can be calculated to have had $(^{228}\text{Th}/^{232}\text{Th}) \sim 5.7 \pm 0.7$. A valid earlier interpretation was that these magmas formed 7 years before eruption.

The 1988 lavas have-compositions consistent with two end-member models. If the magma formed and separated at one instant, then the magma formed 20.4 ± 0.7 years ago. If the magma formed and was immediately transferred to a well-mixed reservoir before eruption, then the magma reservoir has a mean lifetime $\tau = 81 \pm 9$ years and a volume of $1.5 \pm 0.2 \times 10^7\,\text{m}^3$. There are a range of possible intermediate solutions, such that the magma experienced a finite and calculable transfer time t from the source to a reservoir, given by $0 < t < 20$ years, followed by a period of residence τ in a steady-state chamber, given by $81 > \tau > 0$ years. The 1988 magmas experienced a total delay time of 20 to 81 years between segregation and eruption.

1 Introduction

Radiochemical measurements on natrocarbonatite lavas erupted at Oldoinyo Lengai in 1960–1963 and 1988 have been reported elsewhere (Williams et al. 1986; Pyle et al. 1991). These data are summarized below, together with a discussion of their implications for the genesis of carbonatite magma at Oldoinyo Lengai.

---

[1] Department of Earth Sciences, University of Cambridge, Downing Street, Cambridge CB2 3EQ, UK

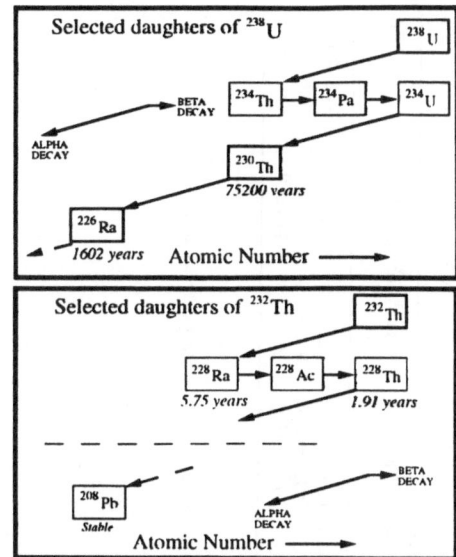

Fig. 1. Summary of the ^{238}U and ^{232}Th decay series. The most useful nuclides, for this case, are shown in *bold boxes*. Figures in *italics* indicate the nuclide half-lives. ^{232}Th and ^{238}U have such long half-lives that they can be regarded as stable on the timescales (10^0–10^2 years) considered here. Many short-lived daughters have been neglected for simplicity. The full decay series can be found in Ivanovich and Harmon (1992)

1.1 U and Th Decay Series Disequilibria

The natural decay series of uranium (^{238}U, ^{235}U) and thorium (^{232}Th) have important applications in the study of recent geological processes (Ivanovich and Harmon 1992). In each series, a long-lived parent passes by a sequence of α- or β-decays through a variety of intermediate daughters to a stable isotope of lead (Fig. 1). The different chemical behaviour of these daughters allows them to be fractionated and hence radioactive disequilibria to be established. The different half-lives of the daughters allow the time elapsed since fractionation to be assessed, on time scales ranging from days to ~350 ka. In this way, disequilibria may be used to constrain the time scales and mechanisms of the genesis of magmas. The theory and application of this method, particularly with reference to the dating and evolution of silicate lavas, has been reviewed recently elsewhere (Condomines et al. 1988; Gill et al. 1992), and will be expanded in the following sections.

Uranium-series disequilibria studies are concerned with the measurement of nuclide decay rates ("activities"), rather than concentrations. The activity A of a nuclide is related to the number of atoms present by $A = \lambda_i N_i$, where λ_i is the decay constant. Thus, nuclides with a high probability of decay (high λ_i) will generally be present only in low concentrations. When a decay series is at secular equilibrium, all species of that series will be decaying at the same rate, so all activity ratios within a given decay series will be equal to 1. Radioactive

Table 1. Selected radiochemical data from Oldoinyo Lengai natrocarbonatites

Eruption year	Notes	Th (ppm)	U (ppm)	^{238}U (dpm/g)	^{230}Th (dpm/g)	^{226}Ra (dpm/g)	$(^{230}\text{Th}/^{232}\text{Th})$	$(^{230}\text{Th}/^{238}\text{U})$	$(^{226}\text{Ra}/^{230}\text{Th})$	$(^{228}\text{Th}/^{232}\text{Th})$	$(^{238}\text{U}/^{232}\text{Th})$
1960[a]	BD 114	1.92 ± 0.05	7.20 ± 0.10	5.37 ± 0.08	0.47 ± 0.01	29.5 ± 0.6	1.01 ± 0.04	0.088 ± 0.003	62 ± 3	3.28 ± 0.17	11.4 ± 0.3
1963[b]	113544	1.17 ± 0.06	6.34 ± 0.12	4.73 ± 0.09	0.32 ± 0.02	24.9 ± 0.5	1.11 ± 0.08	0.067 ± 0.003	78 ± 4	4.18 ± 0.22	16.6 ± 0.9
1988	BD 4152	3.20 ± 0.29	6.27 ± 0.16	4.68 ± 0.11	0.80 ± 0.07	>27	1.03 ± 0.12	0.17 ± 0.02	>34	4.64 ± 0.48[c]	6.00 ± 0.56

[a] Williams et al. (1986).
[b] Pyle et al. (1991).
[c] Mean of five samples is 5.5 ± 0.6. Activites and concentrations were all determined by alpha spectrometry; 1σ errors from counting statistics alone.

disequilibrium refers to any state where the activity ratio of any pair of nuclides from a single decay series is not equal to 1.

1.2 Disequilibria Measurements on Carbonatites

Disequilibria measurements on natrocarbonatites have been determined for lavas erupted between 1960 and 1963 (Williams et al. 1986) and in 1988 (Pyle et al. 1991). These data are summarized in Table 1. The 1960, 1963 and 1988 lavas are all fairly similar compositionally. The lavas show a range of [U/Th] weight ratios, from ~2 to ~6, and have activity ratios ($^{226}Ra/^{230}Th$) of >30 to 78. In the 1988 lavas, ^{228}Th and ^{232}Th are distinctly out of equilibrium, with a mean activity ratio of 5.5 ± 0.6.

2 Long-Lived Disequilibria and Constraints on the Origin of Natrocarbonatite

2.1 Introduction

^{238}U, ^{230}Th, ^{226}Ra and ^{210}Pb are the "longer-lived" nuclides of the ^{238}U decay series. These are highly disequilibrated in fresh natrocarbonatite lava, with substantial enrichment of ^{226}Ra and ^{238}U relative to ^{230}Th, suggesting that Ra, U and Pb all partition preferentially into natrocarbonatite magma. These extreme disequilibria were used by Pyle et al. (1991) to constrain the applicability of different models of carbonatite magmagenesis by mass balance. These calculations were discussed in some detail in Pyle et al. (1991), and are only briefly discussed here.

2.2 Model Assumptions

At the time of carbonatite formation, two phases are assumed to coexist at chemical equilibrium; a carbonatite melt and a (silicate) "conjugate phase". This situation could encompass any of the proposed genetic models for carbonatite, with the "conjugate phase" plausibly representing the crystallising assemblage, unmolten residue or immiscible liquid. Radionuclides partition between the carbonatite and silicate phases, allowing radioactive disequilibria to be established. The whole system is assumed to start at secular equilibrium. With these assumptions, it is possible to write down mass balance equations for each separate nuclide, and to recast these equations in terms of a measurable "partition coefficient" (D[X]) for each nuclide between the two phases, and the observed activity ratios $[R_X = (X_c/^{226}Ra_c)]$ of different nuclide pairs in the carbonatite. In turn this permits calculation of the mass percentage of carbonatite required at equilibrium:

$$\text{Mass\% carbonatite} = \frac{(D[Ra] - D[X] \cdot R_X)}{\{D[Ra] + R_X \cdot (1 - D[X]) - 1\}}. \tag{1}$$

where X = (^{238}U), (^{230}Th) and (^{210}Pb). At Oldoinyo Lengai, the extreme enrichment of ^{226}Ra in the carbonatite suggests that D[Ra] $\approx$ 0, so Eq. (1) simplifies to

$$\text{Mass\% carbonatite} = \frac{D[X].R_X}{\{R_X.(D[X] - 1) + 1\}}. \tag{2}$$

Concentrations of U, Th and Pb in silicate material from Lengai constrain the values of D[U], D[Th], and D[Pb]. Combined with the measured radioactive disequilibria in the 1960, 1963 and 1988 lavas, these data constrain the plausible magma generation models at Oldoinyo Lengai. Permissible solutions (for silicate parents in equilibrium with natrocarbonatite) are those where the required mass percentages of carbonatite for different elements are equal. Calculations for Oldoinyo Lengai samples are detailed elsewhere (Fig. 5 in Pyle et al. 1991). The best agreement between observed disequilibria and D[X] for U, Th and Pb for the 1960 and 1963 lavas is with phonolites or young nephelinites, which require ~10–17 wt% or 4–8 wt% natrocarbonatite, respectively, at equilibrium. The 1988 lavas have disequilibria consistent with the equilibration of 8–17 wt% carbonatite with nephelinite, or 12–21 wt% carbonatite with wollastonite melanephelinite. Nepheline and pyroxene cumulate rocks cannot be in equilibrium with natrocarbonatite of any of the compositions modelled. While there are insufficient partitioning data to properly test fractionation models (Twyman and Gittins 1987; Gittins 1989) the requirement that the fractionating assemblage must have $D_{Th} > D_U$ makes fractional crystallization models an unattractive proposition (Williams et al. 1987).

These calculations confirm that natrocarbonatite could be in equilibrium with silicate magmas of the compositions observed at Oldoinyo Lengai and are consistent with an origin of natrocarbonatite by shallow-level immiscibility of 5–25 wt% carbonatite from a silicate parent melt. Such a hypothesis is consistent with available experimental and petrological studies on natrocarbonatite and related synthetic systems (e.g. Peterson 1990).

3 Short-Lived Disequilibria Constraints on the Timing of Magma Generation

3.1 Introduction

At Oldoinyo Lengai, the disequilibria between the short-lived U and Th decay series nuclides is far in excess of those observed in silicate lavas (Capaldi et al. 1976; Williams et al. 1986; Condomines et al. 1987). In particular, the lavas are both out of ^{228}Th-^{232}Th and ^{228}Ra-^{232}Th equilibrium. Williams and co-workers were the first to document these extreme disequilibria in the products of eruptions from the early 1960s. They argued convincingly that these disequilibria required the lavas to have formed less than 20 years before eruption (Williams et al. 1986); this work gave the impetus for subsequent investigations on more recent lavas from Oldoinyo Lengai.

Two types of model have been used to interpret these short-lived disequilibria in the past (Williams et al. 1986; Pyle et al. 1991). The first model assumes that all of the carbonatite formed at one instant. The observed disequilibria then constrain the timing of this (notional) event. The second model assumes instead that there is a carbonatite magma chamber below Oldoinyo Lengai, which is being continually replenished with fresh magma. In this case, the disequilibria constrain the volume of the chamber as a function of the erupted magma flux. Given suitable model assumptions, these data also constrain the time taken for the magma to rise from the source region to the magma chamber.

3.2 Instantaneous Model

This first model of carbonatite formation assumes that all of the carbonatite separates from a silicate body at one instant. This approximation can be justified, since the chemical diffusivity of components in molten carbonatite is so high that small immiscible droplets will remain equilibrated with the silicate matrix as long as the two are intermixed (Treiman 1989). However, after coagulation of these carbonatite blebs beyond a size of $1-10$ cm, unmixing of natrocarbonatite and silicate magma would occur rapidly, due to the large density difference between the two melts.

The timing of formation can be assessed by comparing $(^{228}Ra/^{232}Th)$ and $(^{226}Ra/^{230}Th)$ at any instant. (^{228}Th) and (^{228}Ra) have half-lives of 1.91 and 5.75 years; (^{226}Ra) and (^{230}Th) are substantially longer-lived. At Oldoinyo Lengai, both $(^{228}Ra/^{232}Th)$ and $(^{226}Ra/^{230}Th)$ are out of equilibrium, indicating that Ra-Th fractionation occurred recently. In fact, the time elapsed since fractionation can be precisely assessed from expressions which describe the evolution of the activity ratios $(^{228}Ra/^{226}Ra)$ and $(^{228}Th/^{232}Th)$, following a fractionation of Ra from Th (Capaldi et al. 1976; Williams et al. 1986; Pyle et al. 1991), viz

$$\left(\frac{^{228}Ra}{^{226}Ra}\right)_t = \left(\frac{^{228}Ra}{^{226}Ra}\right)_0 \cdot e^{(-\lambda_{Ra} \cdot t)} + \left(\frac{^{232}Th}{^{226}Ra}\right)_0 \cdot (1 - e^{(-\lambda_{Ra} \cdot t)}), \tag{3}$$

where λ_{Ra} is the decay constant of ^{228}Ra ($\sim 0.12 a^{-1}$) and

$$\left(\frac{^{228}Th}{^{232}Th}\right)_t = \left(\frac{^{228}Th}{^{232}Th}\right)_0 \cdot e^{(-\lambda_{Th} \cdot t)} + \left(\frac{\lambda_{Th}}{\lambda_{Th} - \lambda_{Ra}}\right) \cdot (e^{(-\lambda_{Ra} \cdot t)} -$$

$$e^{(-\lambda_{Th} \cdot t)}) \cdot \left[\left(\frac{^{226}Ra}{^{230}Th}\right)_0 - 1\right] + [1 - e^{(-\lambda_{Th} \cdot t)}], \tag{4}$$

where λ_{Th} is the decay constant of ^{228}Th.

In the case of "zero-age" samples, simultaneous solution of these equations gives the time t since fractionation, and the initial $(^{226}Ra/^{230}Th)$ activity ratio of the system. For old samples of known age, these equations may be used to correct the activity ratios for decay since the time of eruption. This correction cannot be extended to include $(^{228}Th/^{232}Th)$ unless a further model assumption is made. If an instantaneous formation model is assumed, with the parent magma initially starting at secular equilibrium, then it is possible to calculate the

Table 2. a) $(^{228}\text{Th}/^{232}\text{Th})$ and $(^{228}\text{Ra}/^{226}\text{Ra})$ activities decay corrected to the time of eruption, instantaneous model

Sample	BD 114, 1960	USNM, 1963	1988
$(^{228}\text{Th}/^{232}\text{Th})$	33[c]	36[c]	5.5 ± 0.6[m]
$(^{228}\text{Ra}/^{226}\text{Ra})$	0.44 ± 0.03[d]	0.36 ± 0.03[d]	0.11 ± 0.01[m]
model age on eruption	7 years[c]	8.5 years[c]	20 years[c]

[c] Model-dependent calculation. [m] Measured. [d] Decay-corrected measurement.
Calculations assume that the magma formed at one instant, and then rose to the surface on a timescale referred to as the "model age on eruption".

b) $(^{228}\text{Th}/^{232}\text{Th})$ and $(^{228}\text{Ra}/^{226}\text{Ra})$ activities at the time of eruption, steady-state model assumption 1

Sample	BD 114	USNM	1988
$(^{228}\text{Th}/^{232}\text{Th})$	5.7 ± 0.7[c]	5.6 ± 0.7[c]	5.5 ± 0.6[m]
$(^{228}\text{Ra}/^{226}\text{Ra})$	0.44 ± 0.03[d]	0.36 ± 0.03[d]	0.11 ± 0.01[m]
$(^{226}\text{Ra}/^{230}\text{Th})$	62 ± 3[m]	78 ± 4[m]	51 ± 6[c]
τ	11 ± 1[c]	13 ± 2[c]	81 ± 9[c]

[c] Model-dependent calculation. [m] Measured. [d] Decay-corrected measurement.
Calculations assume that magma is transferred to a homogenised reservoir immediately after formation. For the 1960, 1963 lavas the measured (Ra/Th) and $(^{228}\text{Ra}/^{226}\text{Ra})$ ratios constrain τ [Eq. (5)]; in turn this permits calculation of the model $(^{228}\text{Th}/^{232}\text{Th})$ on eruption from the equivalent of Eq. (5) for Th. For the 1988 lavas, the measured $(^{228}\text{Ra}/^{226}\text{Ra})$, $(^{228}\text{Th}/^{232}\text{Th})$ ratios constrain $(^{226}\text{Ra}/^{230}\text{Th})$.

ratio $(^{228}\text{Th}/^{232}\text{Th})$ on eruption. These calculations are summarized in Table 2. For the instantaneous model the delay between formation and eruption has increased from 7 years in 1960 (Williams et al. 1986) to 20 years in 1988 (Pyle et al. 1991).

3.3 Steady-State Calculations

The same data for the 1960, 1963 and 1988 lavas may be used to model the evolution of the Oldoinyo Lengai magma system, assuming steady state conditions. In this model, a steady influx of magma (rate Q_i) feeds fresh ^{228}Ra into a reservoir of constant volume V. In the reservoir, ^{228}Ra both ingrows from its parent ^{232}Th, and decays; the chamber is continuously erupting magma at a rate Q_o, with $^{228}\text{Ra}_o$ atoms of ^{228}Ra per unit volume. Crystallization is neglected. Q_o is known from field observation to be $500\,\text{m}^3/\text{day}$ in 1988. Applying mass balance (see Pyle 1992), we derive expressions relating the reservoir volume V to the eruption rate Q ($=Q_i = Q_o$) of magma:

$$V = \left\{ \frac{Q \cdot \{(^{228}\text{Ra}/^{226}\text{Ra})_{\text{in}} - (^{228}\text{Ra}/^{226}\text{Ra})_{\text{out}}\}}{\lambda_{228\text{Ra}} \cdot [r - (^{232}\text{Th}/^{226}\text{Ra})]} \right\}, \tag{5}$$

where r is the $(^{228}Ra/^{226}Ra)$ activity ratio in the chamber, λ_{228Ra} is the decay constant of ^{228}Ra, $(^{228}Ra/^{226}Ra)_{in}$ is the activity ratio of the magma entering the chamber and $(^{228}Ra/^{226}Ra)_{out}$ is the activity ratio of the material exiting the chamber. The freshly formed natrocarbonatite entering the chamber will have $(^{228}Ra/^{226}Ra)_{in} \sim 1.0$, since $(^{228}Ra/^{226}Ra) = (^{232}Th/^{230}Th)$ at the instant of Ra-Th fractionation (Table 1). The reservoir ratio r is constrained by $(^{228}Ra/^{226}Ra)_{out} < r < (^{228}Ra/^{226}Ra)_{in}$. The residence time τ of a component in an open magma chamber is given by $\tau = (V/Q)$. So, clearly, by dividing Eq. (5) by Q we may obtain expressions for τ. The radium data alone for the 1988 lavas at Oldoinyo Lengai indicate that $\tau = 81 \pm 9$ years (Pyle et al. 1991).

The same steady-state calculations can be performed for $(^{228}Th/^{232}Th)$, and indeed for any other activity ratio (Pyle 1992). At Oldoinyo Lengai, the $(^{228}Th/^{232}Th)$ data alone do not help to constrain τ, due to uncertainty in the $(^{228}Th/^{232}Th)$ ratio of the incoming magma. Depending on the time elapsed since "formation" and the $(^{226}Ra/^{230}Ra)$ activity ratio of the freshly formed carbonatite, the ratio $(^{228}Th/^{232}Th)$ could range from 1–60. To overcome this, one can make either of two assumptions, discussed below.

3.3.1 Model Assumption 1: Negligible Magma Transfer Time from the Source to the Chamber

If the magma supplied to the chamber is assumed to have formed a very short time–on the order of days or weeks–before entering the chamber, then the Ra and Th activity ratios of this influent magma must be given by $(^{228}Ra/^{226}Ra)_{in} = (^{228}Th/^{232}Th)_{in} = 1.0$. For the 1988 lavas, the conditions that $(^{228}Ra/^{226}Ra)_{chamber} = 0.11$, and $(^{228}Th/^{232}Th)_{chamber} = 5.5$ are satisfied when the chamber activity ratio $(^{226}Ra/^{230}Th) = 51 \pm 6$, the volume of the magma chamber is $1.5 \pm 0.2 \times 10^7 \, m^3$ and $\tau = 81 \pm 9$ years. For the 1960 and 1963 lavas, the same assumption, combined with the measured $(^{226}Ra/^{230}Th)$ activity ratio of the lavas constrains the model value of $(^{228}Th/^{232}Th)_{chamber}$ in 1960 and 1963 to be ~ 5.7. These calculations are all summarised in Table 2.

3.3.2 Model Assumption 2: Finite Magma Transfer Time from the Source to the Chamber

A second series of solutions can be obtained by making no assumption about the initial compositions of the influent magma. Instead, we assume that the plumbing system at Oldoinyo Lengai comprises a source region, where freshly disequilibrated magma is continually produced, and a steady-state magma reservoir. Magma may take a finite period of time, the transfer time, to travel from the source region to the magma chamber. The compositions of the erupted magmas are assumed to be $(^{228}Th/^{232}Th) = 5.5$, $(^{228}Ra/^{226}Ra) = 0.11$. By equating expressions of the form of Eq. (5) for $(^{228}Ra/^{226}Ra)$ and $(^{228}Th/^{232}Th)$, since V calculated from each set of data must be equal, we can obtain a relationship between the ratios $(^{228}Ra/^{226}Ra)_{in}$ and $(^{228}Th/^{232}Th)_{in}$ of the influent magma.

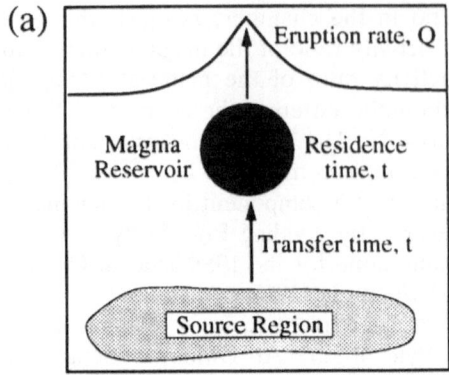

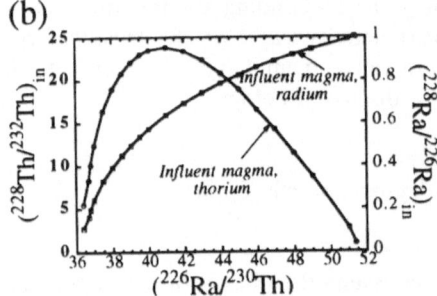

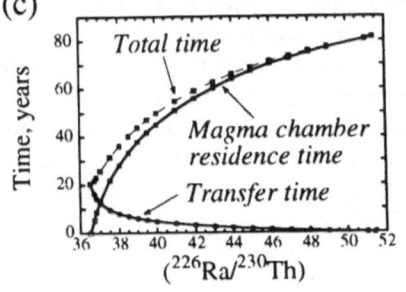

Fig. 2a–c. Summary of the model transfer and storage time calculations. **a** Models assume that the plumbing system at Oldoinyo Lengai comprises a source region, where freshly disequilibrated magma is continually produced, and a steady-state magma reservoir. Magma may take a finite period of time to travel from the source region to the magma chamber – this is called the transfer time. The compositions of the erupted magmas are assumed to be $(^{228}Th/^{232}Th) = 5.5$, $(^{228}Ra/^{226}Ra) = 0.11$, and the $(^{226}Ra/^{230}Th)$ ratio, which is only constrained by measurement be >34, is assumed unaffected by decay over the short period of time between the source region and eruption. **b** Summary of the calculations of the composition of the magma entering the steady-state reservoir. The curve *influent magma, radium* refers to the model $(^{228}Ra/^{226}Ra)$ of the incoming magma as a function of the ratio $(^{226}Ra/^{230}Th)$ of the system. The other curve refers to the $(^{228}Th/^{232}Th)$ ratio of the same magma. These are uniquely constrained as a function of $(^{226}Ra/^{230}Th)$. **c** Summary of the time elapsed between magma generation in the source region and entry into the magma chamber. The transfer time is required to allow $(^{228}Ra/^{226}Ra)$ and $(^{228}Th/^{232}Th)$ ratios to decay to a value which will generate the observed Ra and Th activity ratios in the steady-state magma chamber

This varies as a function of the ratio $(^{226}Ra/^{230}Th)$ in the magma, which has not been determined for the 1988 lavas, and depends on the time elapsed between formation and arrival at the reservoir.

From earlier [Eqs. (3,4)], the evolution of the carbonatite magma composition following instantaneous separation from the magma source region is also known. Thus for any batch of magma rising from the source to the chamber it is possible to compute the transfer time from source to reservoir, as well as the storage time in the reservoir. This has been done for the 1988 lava composition, with the results shown graphically in Fig. 2b and c. Figure 2b shows the $(^{228}Ra/^{226}Ra)$ and $(^{228}Th/^{232}Th)$ compositions of the magma entering the steady-state reservoir. These are uniquely constrained as a function of $(^{226}Ra/^{230}Th)$, and represent the range of possible solutions which will produce the observed disequilibria at the surface, after a measurable period of residence in the magma reservoir. Figure 2c summarizes the time elapsed between magma generation in the source region and entry into the magma chamber "transfer time" together with the magma chamber residence time (τ), and the total time elapsed between formation and eruption (which is just the sum of the transfer time and τ). It is clear from this that at the limits these calculations coincide with the instantaneous model (when $\tau = 0$), or with the steady-state model, when the transfer time is 0. The same calculations cannot be performed for the 1960, 1963 lavas since the $(^{228}Th/^{232}Th)$ of the magma chamber is unknown.

In conclusion, if a short transfer period is allowed, so that the magma can rise from the source region to the magma chamber, then this period is between 0 and $20(\pm 1)$ years, and the concomitant residence time τ in the chamber is between 81 and 0 years. In any case, the total time between carbonatite segregation and eruption must be between 20 and 81 years. The method developed here is completely general, and can be applied to any other silicate volcano where the steady-state approximation is valid. The model developed here demonstrates that there are a range of possible interpretations for the time taken between magma formation and eruption. With the available radiochemical data, it is not possible to distinguish between an "instantaneous" formation model, where there is no storage before eruption, or a "steady-state" model where there is a long period of storage before eruption. A most likely solution is that an intermediate model is appropriate, encompassing a finite rise time and a finite period of storage; this is at least consistent with the observations that Oldoinyo Lengai has been in semi-continuous eruption for the past decade, and with observations on the degassing behaviour of Oldoinyo Lengai in 1988 (Pyle et al., Chap. 3, this Vol.).

3.4 Implications for the Crystallization History of Oldoinyo Lengai

Cashman and Marsh (1988) have shown how the textural characteristics of phenocrysts and their size distributions in lavas can be quantitatively assessed and used to interpret the growth rates of crystals in a steady-state magma chamber. The theory has been discussed at length elsewhere (Marsh 1988; Randolph and Larson 1988). Briefly, in a steady-state system in which crystals

are continuously growing in a recharged magma chamber the population balance
may be described by

$$\frac{dn}{dL} + \frac{n}{G\tau} = 0, \tag{6}$$

where, as before, $\tau = V/Q$, and n is the number of objects of size L, and
growing at a linear rate G. Solving this gives

$$n = n_o \cdot \exp\left[-\frac{L}{G\tau}\right], \tag{7}$$

where n_o is the nucleation density at $L = 0$. Thus, on a plot of $\log_e(n)$ against L,
one should obtain a straight line of slope $-1/G\tau$, and an intercept of n_o. This
has in fact proved to be the case for crystal size distributions (CSDs) in lavas
(Cashman 1988; Cashman and Marsh 1988).

Peterson recently performed the same measurements for a sample of 1963
natrocarbonatite lava from Oldoinyo Lengai (Peterson 1990). He found approximately linear, and similar CSDs for nyerereite and gregoryite phenocrysts from
this lava. The slopes of the $\log_e(n)$-L plot indicated respective values for the
product of crystal growth rate and residence time, $G\tau$, of 0.13 and 0.16 mm. This
value is significantly different from $G\tau$ measured on silicate lavas. For example,
a block of combeite nephelinite ejected from Oldoinyo Lengai during the 1966
eruption has $G\tau \sim 0.03$ mm (Peterson 1990). To interpret these data, Peterson
adopted a residence time of 2 years for these lavas. In fact, as shown here, a
more realistic value for 1963 is 13 ± 2 years (Table 2). This in turn gives a mean
linear growth rate for both phases of 11 ± 2 μm per year, or $3.6 \pm 0.6 \times
10^{-7}$ μm s^{-1}. This value is of the same order as the growth rate of plagioclase
crystals in the Makaopuhi lava lake on Hawaii (5.4–9.9×10^{-7} μm s^{-1}) and in
the pre-1980 dome at Mount St. Helens (3–10×10^{-7} μm s^{-1}, plagioclase)
(Cashman 1988; Cashman and Marsh 1988). Thus, there is no evidence that
carbonate phenocrysts grow any faster than silicate phenocrysts given similar
conditions. The similarity in growth rates between Oldoinyo Lengai natrocarbonatite and those inferred for Mount St. Helens probably reflect the fact that
the systems are cooling at the same rate. Oldoinyo Lengai is the only place
where both $G\tau$ and τ have been quantitatively measured or calculated. There
are as yet no quantitative CSD data on the 1988 lavas for comparison.

4 Summary

Short-lived radioactive disequilibria measured in the natrocarbonatite lavas of
Oldoinyo Lengai provide several unique constraints on the timing and mechanism
of magmagenesis.

Disequilibria between ^{238}U-^{230}Th-^{226}Ra-^{210}Pb are consistent with an origin of
natrocarbonatite by immiscibility. Depending on the composition of the parental
magma, between 5 and 25 wt% of this exsolved to form natrocarbonatite. The

particular parent silicate magma cannot be definitively constrained; it could be a peralkaline nephelinite, or intermediate between nephelinite and phonolite.

^{232}Th-^{228}Ra-^{228}Th disequilibria constrain the time scale of magma generation, ascent and evolution. The available data are consistent with the existence of a steady-state constant volume magma reservoir below Oldoinyo Lengai. The 1960 and 1963 lava compositions indicate that this reservoir had a mean lifetime τ of 11 ± 1 to 13 ± 2 years. For the 1988 lavas, data are consistent with two end-member models. If there is no reservoir and $\tau = 0$, then the magma formed 20.4 ± 0.7 years ago, and has $(^{226}Ra/^{230}Th) = 36$. If the transfer time from the magma source to the magma reservoir is negligible, then the magma reservoir has a mean lifetime $\tau = 81 \pm 9$ years, a volume of $1.5 \pm 0.2 \times 10^7 \, m^3$ and $(^{226}Ra/^{230}Th) = 51$. For $(^{226}Ra/^{230}Th)$ between 36 and 51, the magma experienced a finite and calculable transfer time t, given by $0 < t < 20$ years, followed by a period of residence τ in a steady-state chamber, given by $81 > \tau > 0$ years. The magmas erupting in 1988 experienced a total delay time of 20 to 81 years between segregation and eruption.

Acknowledgements. Some of this work formed a portion of the author's PhD thesis, supported by NERC, and was completed while a Visiting Associate at the California Institute of Technology and Research Fellow at St. Catharine's College, Cambridge. I am very grateful to Steve Sparks, Miro Ivanovich and Barry Dawson for discussion and encouragement over the past few years, and to Ed Stolper and the Experimental Petrology group at Caltech for their hospitality. Critical reviews by Jim Gill, Michel Condomines and Gail Mahood were appreciated. Department of Earth Sciences Contribution Number 2530.

References

Capaldi G, Cortini M, Gasparini P, Pece R (1976) Short lived radioactive disequilibria in freshly erupted volcanic rocks and their implication for the preeruption history of a magma. J Geophys Res 81:350–358

Cashman KV (1988) Crystallisation of Mount St. Helens 1980–1986 dacite: a quantitative textural approach. Bull Volcanol 50:194–209

Cashman KV, Marsh BD (1988) Crystal size distribution (CSD) in rocks and the kinetics and dynamics of crystallization II: Makaopuhi lava lake. Contrib Mineral Petrol 99:292–305

Condomines M, Bouchez R, Ma JL, Tanguy JC, Amosse J, Piboule M (1987) Short-lived radioactive disequilibria and magma dynamics in Etna volcano. Nature 325:607–609

Condomines M, Hemond C, Allègre CJ (1988) U-Th-Ra radioactive disequilibria and magmatic processes. Earth Planet Sci Lett 90:243–262

Dawson JB (1962) The geology of Oldoinyo Lengai. Bull Volcanol 24:349–387

Dawson JB, Bowden O, Clark GC (1968) Activity of the carbonatite volcano, Oldoinyo Lengai. Geol Rundsch 57:865–879

Donaldson CH, Dawson JB, Kanaris-Sotiriou R, Batchelor RA, Walsh JN (1987) The silicate lavas of Oldoinyo Lengai. Neues Jahrb Min Abh 156:247–279

Gill JB, Pyle DM, Williams RW (1992) Igneous Rocks. In: Ivanovich M, Harmon RS (eds) Uranium series disequilibrium: applications to earth, marine and environmental sciences. Oxford University Press, Oxford, pp 207–258

Gittins J (1989) The origin and evolution of carbonatite magmas. In: Bell K (ed) Carbonatites – genesis and evolution. Unwin Hyman, London, pp 580–600

Ivanovich M, Harmon RS (eds) (1992) Uranium-series disequilibrium: applications to earth, marine and environmental sciences. Oxford University Press, Oxford

Marsh BD (1988) Crystal Size Distribution (CSD) in rocks and the kinetics and dynamics of crystallisation I. Theory. Contrib Mineral Petrol 99:277–291

Peterson TD (1990) Petrology and genesis of natrocarbonatite. Contrib Mineral Petrol 105: 143–155

Pyle DM (1992) The volume and residence time of magma beneath active volcanoes determined by decay-series disequilibria methods. Earth Planet Sci Lett 112:61–73

Pyle DM, Dawson JB, Ivanovich M (1991) Short-lived decay series disequilibria in the natrocarbonatite lavas of Oldoinyo Lengai, Tanzania: constraints on the timing of magma genesis. Earth Planet Sci Lett 105:378–396

Randolph, AD, Larson MA (1988) Theory of particulate processes: analysis and techniques of continuous crystallisation, 2nd edn. Academic Press, New York

Treiman AH (1989) Carbonatite magma: properties and processes. In: Bell K (ed) Carbonatites – genesis and evolution. Unwin Hyman, London, pp 89–104

Twyman JD, Gittins J (1987) Alkalic carbonatite magmas: parental or derivative? In: Fitton JG, Upton BGJ (eds) Alkaline igneous rocks. The Geological Society of London, London, pp 85–94

Williams RW, Gill JB, Bruland KW (1986) Ra-Th disequilibria systematics: timescale of carbonatite magma formation at Oldoinyo Lengai volcano, Tanzania. Geochim Cosmochim Acta 50:1249–1259

Williams RW, Gill JB, Bruland KW (1987) Ra-Th disequilibria: timescale of carbonatite magma formation at Oldoinyo Lengai volcano, Tanzania (Reply). Geochim Cosmochim Acta 52:939

An Assessment of the Alleged Role of Evaporites and Saline Brines in the Origin of Natrocarbonatite

K. Bell[1] and J.B. Dawson[2]

Abstract

The high Na content of natrocarbonatite from Oldoinyo Lengai can be explained by extreme differentiation of possible mantle-derived parent magmas. There is no need to elevate the Na content by dissolution of bedded evaporites or interaction with saline brines. The stable and radiogenic isotope signatures of the natrocarbonatites are typical of magmatic rocks and differ significantly from those of trona deposits. Nd and Sr isotopic compositions of natrocarbonatites from Oldoinyo Lengai are similar to other young carbonatites from East Africa and fall on the East African Carbonatite Line defined by initial $^{143}Nd/^{144}Nd$ and $^{87}Sr/^{86}Sr$ ratios. Trace-element signatures of natrocarbonatite are similar to those of other carbonatites, and their lineage is clearly magmatic. If natrocarbonatites were either remobilized or dissolved trona deposits, the resulting melt would have to acquire almost all of the chemical features observed in known carbonatites from East Africa.

1 Introduction

The sodium carbonate lavas (natrocarbonatites) of Oldoinyo Lengai are unique among carbonatites and, because of this, some researchers have felt that these unusual rocks deserve a somewhat unusual origin. The natrocarbonatites are unquestionably distinct and differ from other carbonatites in their high alkali content and unusual mineralogy. Combined total alkalies are about 40 wt%, (Na_2O making up about 32%), and the Na_2O/K_2O ratio is about 4. The unusual nature of the Oldoinyo Lengai natrocarbonatites, coupled with the fact that they are the only samples that have been collected prior to any weathering or exposure to meteoric waters, has led many workers to suggest that this magma type should occupy a prominent role in carbonatite genesis and has led to the suggestion that many carbonatites magmas were once alkali-rich (Dawson 1962, 1964; Woolley 1969; Le Bas 1989). The absence of alkalies in most other carbonatites has been attributed either to loss during fenitization, in which

[1] Ottawa-Carleton Geoscience Centre, Department of Earth Sciences, Carleton University, Ottawa, Canada K1S 5B6
[2] Grant Institute of Geology, University of Edinburgh, West Mains Road, Edinburgh EH9 3JW, UK

alkalis are transported from the carbonatite into the surrounding country rocks by fluids, or to alkali loss during recrystallization.

The abundance of alkalies in natrocarbonatites has been attributed to: (1) enrichment during crystal fractionation of a primary olivine sövite (Gittins 1989), (2) the production of conjugate liquids by immiscibility (Kjarsgaard and Hamilton 1989), (3) differentiation of a carbonated silicate melt by both crystal fractionation and liquid immiscibility (Le Bas 1989), and (4) interaction between a silicate magma and either soda-rich evaporite deposits (Milton 1968, 1989) or saline brines (Eugster 1970). The first three, although differing, assume an igneous parentage, and it is the purpose of the present chapter to evaluate the last of these four models.

Milton (1968) initially proposed that the natrocarbonatite from Oldoinyo Lengai had a surficial origin resulting from reaction between local troniferous lacustrine sediments and a nephelinitic magma. He argued, on the basis of phase equilibrium studies, that a rock chemically and mineralogically similar to a natrocarbonatite "should crystallize as a peri-eutectic from a complex troniferous sediment-alkalic magma system; and should separate out as an immiscible carbonate melt from a heavier silicate fluid". Milton (1989) cited, in favour of this eutectic composition, the system $CaCO_3$-Na_2CO_3 worked on by Niggli (1916). Evidence used by Milton (1989) to support his model included (1) the presence of voluminous trona deposits not too far distant from Oldoinyo Lengai, (2) the absence in natrocarbonatites of many of the silicate or "rare" minerals present in ordinary carbonatites, and (3) the unusual chemical composition of natrocarbonatite. Eugster's interest in the source of the salts in evaporite deposits of the East African rift, particularly those of Lake Magadi, led him to become peripherally interested in the natrocarbonatite problem. Of interest was whether the Magadi trona could ultimately be derived from the Lengai eruptions. Eugster (1970) argued that direct surface leaching of the Oldoinyo Lengai natrocarbonatite lavas could be eliminated because of the hydrologic separation of the Natron and Magadi basins, and ruled out wind-borne ash on the grounds that Lake Natron should be the trona-rich lake, rather than Magadi, since it lies closer to Oldoinyo Lengai. A model was proposed that involved a deep, saline groundwater reservoir for the Magadi basin and also for the Natron springs. To quote, "the sodium-carbonate lavas of Lengai then become not so much remobilized trona, but simply the product of the interaction between the alkaline lavas and this groundwater reservoir". Eugster (1970, 1986), therefore, presented a similar, but alternative, model to that of Milton, explaining the natrocarbonatite lava and ash of Lengai in terms of interaction of an "alkaline silicate magma with a saline alkaline groundwater reservoir". In 1986 the controversy was renewed by Peterson and Marsh, who stated ". . . it is probable that the extreme composition of Oldoinyo Lengai is due to assimilation of trona, with subsequent exsolution of natrocarbonatite". This was subsequently rejected by Peterson (1990).

Little data-based argument has been published about the two models involving either trona or brine. Although somewhat ignored by many interested in the generation of natrocarbonatite and carbonatites, the models deserve a public airing coupled with critical evaluation. A critical assessment is long overdue.

2 Discussion

The close proximity of Oldoinyo Lengai to evaporite deposits, particularly to Lake Natron with its abundant supply of sodium carbonate, is beyond dispute. The Natron and Magadi lagoons are fed by both sheet floods from ephemeral streams, and by perennial springs (commonly thermal) flowing over and through the surrounding volcanic terrains. Each season a trona layer, 2–5 cm thick forms; the average accumulation rate during the Holocene at Lake Magadi is 0.4 cm/a. At Lake Natron, trona crusts are as much as 7 m thick (Guest and Stevens 1951). Eugster (1980) estimated that approximately 75% of the trona in the lakes is recirculated each year by dissolution and reprecipitation. Although alkali evaporites, rich in sodium, are abundant near Oldoinyo Lengai, the questions remain as to whether a silicate magma reacted with them and, if so, what was the result of this reaction?

2.1 Stratigraphy

Milton, in the past, contended that natrocarbonatite is derived from trona-gaylussite-pirsonnite beds similar to those that occur near the volcano and which, he claimed, may underlie the volcano. There is little evidence that lacustrine deposits do underlie the volcano. In fact, Oldoinyo Lengai, with the neighbouring Kerimasi volcano, erupted on a structurally and topographically high basalt ridge separating the Lake Natron basin from the Engaruka Basin. To the east and south of Oldoinyo Lengai are explosion craters in which Oldoinyo Lengai pyroclastic deposits can be seen overlying older basalts and pyroclastics associated with Kerimasi (Dawson and Powell 1969). There is no sign of evaporite beds in any of these craters or even lacustrine sediments similar to the Peninj Beds, which were deposited in a westerly extension of Lake Natron in mid-Pleistocene times (Isaac 1965; Hay 1966). In short, the evidence is that Oldoinyo Lengai erupted on a basalt/tephra ridge, not through a basin infilled by continental evaporite and clastic deposits.

It could be argued that little is known about the sedimentary succession that existed during the eruptive history of Oldoinyo Lengai, and there is not a great deal of evidence that could be used to rebut such an argument. However, it is surprising that in the East Africa rift, where closed continental basins and evaporites abound, no other carbonatite similar to the natrocarbonatite flows of Oldoinyo Lengai has been found. Carbonatites from nearby Kerimasi (Mariano and Roeder 1983) are clearly calcite-rich flows with zoned primary magmatic calcite phenocrysts, and carbonatites from Shombole (Peterson 1989) are similarly calcium-rich. It seems surprising that in the presence of all of the essential ingredients for the "evaporite" hypothesis, no other sodium-rich carbonatites are associated with volcanoes erupted elsewhere in Kenya and Tanzania, although it could also be argued that the alkalies are lost during fenitization.

2.2 Mineralogy

Recent work by Dawson et al. (1990) and by Keller and Krafft (1990) has shown that natrocarbonatites are much more complicated, in terms of their mineralogy, than was previously thought. In addition to phenocrysts of nyerereite and gregoryite, minerals recognized in the matrix include Na-sylvite, fluorite and Fe-alabandite, with rarer witherite, apatite and sellaite (MgF_2). Dawson et al. (Chap. 4, this Vol.) highlight the fact that many of the phases are unusual high-temperature solid-solutions. Few of these minerals are found in other carbonatites, and silicate minerals present in many carbonatites have yet to be found in natrocarbonatite.

There is general agreement that a natrocarbonatite melt with such a low MgO content (<1%) and high alkalis cannot be produced by direct partial melting of a mantle source, and that either crystal fractionation or liquid immiscibility has to be invoked in order to generate the composition of natrocarbonatites. Evidence is available that supports both methods of magma differentiation. Gittins (1989) argues that a halogen-rich melt, low in P, would generate an alkali-rich differentiate, whereas Kjarsgaard and Hamilton (1989) have shown that immiscibility can occur over a wide range of starting compositions for a silicate parental liquid, although there is still discussion about the pressures at which such a process can take place. The high halogen content may, however, control both the liquidus temperatures and viscosity (Gittins 1991), such that any early-formed minerals may have been removed as cumulate phases and separated from residual natrocarbonatites. The high F content of Lengai may simply reflect a much greater residence time for the parental magma at crustal levels than other parental melts, and more extreme fractional crystallization than encountered in other carbonate melts. The latter may explain why so few of the minerals characteristic of plutonic carbonatites are encountered. Current data show that a strongly peralkaline melt can coexist immiscibly with a natro-carbonatite (Kjarsgaard et al., Chap. 12, this Vol.), at very low pressures.

Some interesting new findings that directly relate to the trona assimilation model are starting to emerge. Recent experiments by Kjarsgaard (pers. comm. 1993) involving mixtures of phonolite + trona and nephelinite + trona in the proportions 70:30 and 85:15, have now shown that at low pressures and temperatures (<1 kb and 725 °C) a sodium-rich carbonate liquid exsolves from the mixtures. Although the chemical composition of this carbonate liquid is grossly similar to natrocarbonatite, in detail there are differences. These include a lower K/Na ratio in the natrocarbonatites reflecting the nil-K composition of the experimental carbonate, as well as higher Fe/Mn ratios and considerably less halogens. An interesting point, however, is the transfer of K from the conjugate silicate melt during the experiments. Although these and earlier laboratory experiments have established that both natrocarbonatite and trona are miscible with a silicate melt and that an immiscible carbonate can exsolve, this in itself does not automatically mean that such a process was involved in the generation of natrocarbonatite at Oldoinyo Lengai. The fact that there are experimentally proven ways to produce an alkali-rich carbonatite from a carbonated-silicate melt must weaken the alternative model that involves contamination of a silicate melt by evaporites.

2.3 Composition

The unusual chemical characteristic of the Oldoinyo Lengai carbonate flows is the high Na_2O content, which in the "lake evaporite-brine" hypotheses is seemingly explained by an unusual mechanism. There are many silicate rocks at Oldoinyo Lengai, such as nephelinites and phonolites, that are themselves rich in sodium. In Dawson's (1989) summary of eruptive activity at Oldoinyo Lengai, the Na_2O contents of the nephelinites, phonolitic nephelinites and phonolites range between 8.54 and 13.53 wt%. Sufficient alkalies are associated with the silicate lavas, and there is no need to resort to an additional source, such as evaporites. Nephelinites are found in both oceanic and continental settings, and all have Na_2O abundances at the percent levels and few, if any, are associated with evaporites. However, natrocarbonatites have yet to be found associated with other nephelinitic centres.

Another element that is critical in assessing the "lake evaporite" hypothesis appears to be potassium, which comprises 5–7% of the bulk total of natrocarbonatite. No potassium salts are known in Lake Natron and although waters from Lake Natron and Magadi contain some potassium, in none of the analyzed waters does the Na_2O/K_2O ratio approximate the values (3.5–4.0) found in natrocarbonatite. The only major source of potassium in the Lake Natron sediments is the shales and, if these had been the source of the potassium in the natrocarbonatite, their involvement might be expected to result in the incorporation of silica and alumina also. The negligible amounts of silica and alumina in natrocarbonatite would seem to preclude the derivation of the potassium from the Natron sediments, although this absence could be attributed to partitioning during immiscibility into a silicate conjugate liquid. On the other hand, appreciable amounts of potassium are present in the silicate rocks of Oldoinyo Lengai, the nepheline containing 20 to 30% of the kalsilite molecule.

The trace element assemblages of the natrocarbonatite are comparable to those found in other carbonatites except for the halogen content. Enrichment of Sr, Ba, REE, Th, Pb and Zr are characteristic of carbonatites (Barker 1989), coupled with enrichment in the light REE (Woolley and Kempe 1989). The exceptions are Nb, Zr and Y, which in natrocarbonatite are either absent or present in only negligible quantities. Woolley's recent compilation of the average chemical compositions of calciocarbonatites, magnesiocarbonatites and ferrocarbonatites (Woolley and Kempe 1989) shows the extreme range of Nb, Zr and Th abundances; thus the absence or the low concentrations of these elements does not automatically preclude "natrocarbonatite" from being a carbonatite. In fact, many carbonatites, such as the Dorowa and Shawa carbonatites of Zimbabwe, most South African carbonatites and many from Canada, contain negligible amounts of Nb. The low Y content (<13 ppm; Keller and Krafft 1990) of the June, 1988, eruption is puzzling but, in contrast, the Ba, Sr and V contents are comparable to many carbonatites.

2.4 Isotopic Signatures

The isotope ratios of Nd, Sr, O and C in natrocarbonatite all fall within the ranges exhibited by other carbonatites (Bell et al. 1973; Bell and Blenkinsop 1987; Keller and Krafft 1990; Bell and Dawson, Chap. 7, this Vol.; Dawson et al., Chap. 4, this Vol.). Of particular interest are the Nd and Sr isotope data that fall close to primitive, undifferentiated planetary values and lie on the linear array established by Bell and Blenkinsop (1987) for other recent carbonatites from East Africa. If the "evaporite" model were accepted, then the evaporite would have to completely dissolve in the silicate melt, to be followed by immiscibility, which led to the formation of natrocarbonatite that inherits the isotopic characteristics of the host magma.

The distinct C and O isotopic differences that exist between trona and natrocarbonatite are shown in Fig. 1. Three fields are shown, one for sedimentary

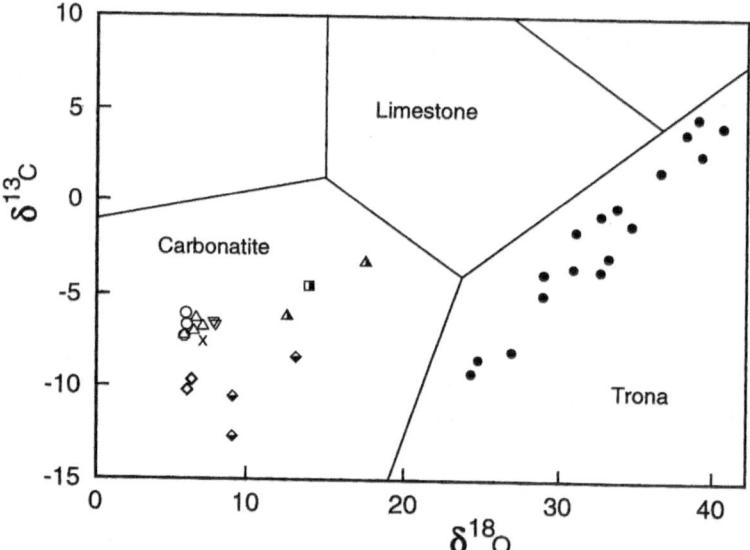

Fig. 1. Plot of $\delta^{13}C$ (PDB) and $\delta^{18}O$ (SMOW) from natrocarbonatites and trona. The limestone and carbonatite fields are based on published data summarized by Bell and Card (in prep.). Limestone field based on 74 individual limestone units and the carbonatite field on data points from 57 complexes. Compilation based on 956 samples. Sources of Oldoinyo Lengai data: *triangles* (Keller and Hoefs, Chap. 8, this Vol.); *diamonds* (Suwa et al. 1975); *inverted triangles* (Dawson 1989); *square* (Denaeyer 1970); *cross* (O'Neil and Hay 1973); *open circles* (Dawson et al., Chap. 4, this Vol.). *Half shaded symbols* mark data from natrocarbonatite affected by post-crystallization processes. *Dots* represent data from trona. (O'Neil and Hay 1973; Suwa et al. 1975)

limestones, one for known carbonatites and one for trona. The carbonatite field includes both altered and unaltered natrocarbonatites from Oldoinyo Lengai. Five samples of natrocarbonatite from the 1960 eruption yielded $\delta^{18}O$ values of 6.2 to 6.4‰, and $\delta^{13}C$ values of -12.6 to -8.4‰ for the inner parts of the specimens, whereas the outer part and crushed samples had significantly higher $\delta^{18}O$ values of 9.0 to 13.1‰ (Suwa et al. 1975). Other investigators have also reported values for natrocarbonatite of -7.6 to -4.5‰ for $\delta^{13}C$ (Denaeyer 1970; Vinogradov et al. 1970; O'Neil and Hay 1973; Sheppard and Dawson 1973; Dawson et al. 1990), which are higher than those reported by Suwa et al. (1975), and $\delta^{18}O$ of $+5.9$ to $+7.1$‰ (O'Neil and Hay 1973; Dawson et al. 1990). The new data from the 1988 eruption (Dawson et al., Chap. 4, this Vol.; Keller and Hoefs, Chap. 8, this Vol.) of unaltered whole-rock samples and their nyerereite and gregoryite phenocrysts lie within fairly restricted ranges of $\delta^{13}C$ -6.3 to -7.1 and 5.8 to 6.7 for $\delta^{18}O$. Trona samples from Lake Magadi (O'Neil and Hay 1973; Suwa et al. 1975) yield quite different results to those found in the natrocarbonatites. The $\delta^{18}O$ values range from 24.4 to 39.2‰, whereas the $\delta^{13}C$ values range from -9.0 to $+2.8$‰. Vinogradov et al. (1970) reported a $\delta^{13}C$ of -0.9‰ for trona crust and $+2.0$‰ for travertine from Lake Magadi. Such relatively high values for $\delta^{18}O$ relative to the natrocarbonatite are in keeping with trona being an evaporite mineral. Overall, trona has much higher $\delta^{18}O$ values for a given $\delta^{13}C$ value. All carbonatites and most trona deposits that have been measured have $\delta^{13}C$ values of zero or less, whereas all trona deposits have $\delta^{18}O$ values greater than $+20$. Carbonatites with $\delta^{18}O > 20$‰ (e.g. $\delta^{18}O$ of 24.5 for altered natrocarbonatite from Oldoinyo Lengai; Dawson 1993) are either altered or have recrystallized in the presence of meteoric water.

So far we have not touched on the question of what happens to the $\delta^{18}O$ and $\delta^{13}C$ values during mixing of a silicate melt and trona. Mixing a nephelinitic magma and a Na-rich evaporite or brine could theoretically yield an isotope signature different from any of the values that characterize both end-members, the extent and the direction of change depending on the abundances of C and O in both end-members, and the temperature at which assimilation takes place. Evolution of CO_2 gas during volcanism is known to fractionate heavy carbon into the gas phase concentrating the lighter isotopes in the melt. At temperatures between 300 and 800 °C carbon in $CaCO_3$, co-existing with CO_2 gas, is lighter than carbon in the CO_2 gas by 2–3 per mill (Bottinga 1968). The isotope composition of the CO_2 gas phase from Oldoinyo Lengai is -2.62‰ $\delta^{13}C$ and $+12.9$ to 18.9‰ $\delta^{18}O$ (Javoy et al. 1989).

Without any experimental data it is difficult to assess the magnitude and direction of the isotopic shifts that could be brought about by exchange reactions and isotopic fractionation in the system nephelinite-trona. The mere fact, however, that the C and O values of natrocarbonatite are similar to those of many other carbonatites suggests that there may have been either little involvement of outside sources, such as would be expected by either contamination or assimilation, or that the isotopic signature of the silicate melt has dominated the final product produced by assimilation, or that decarbonation has lightened to C and O that has been contributed by the trona. However, it seems highly unlikely that during any interaction the isotopes of Nd, Sr, C and O, of widely differing masses, should have re-equilibrated to give the ratios characteristic of other carbonatites.

2.5 Disequilibrium Data

In 1986, Williams et al. published their measurements from natrocarbonatite from Oldoinyo Lengai and reported huge $^{238}U/^{230}Th$ and $^{226}Ra/^{230}Th$ disequilibria, the largest ever measured in any volcanic rock. Carbonatite had $^{238}U\text{-}^{230}Th$ activity ratios from 11.3 to 15.0, and $^{226}Ra/^{230}Th$ activity ratios from 62.8 to 83.0. Ash had a $^{238}U/^{230}Th$ ratio of 1.6 and a $^{226}Ra/^{230}Th$ ratio of 2.07. The $^{230}Th/^{232}Th$ ratios of all measured samples were close to unity. Disequilibrium was also shown by the $^{228}Ra/^{232}Th$ ratio from natrocarbonatite of about 27, representing the only unequivocal example of excess ^{228}Ra in young volcanic rocks (Condomines et al. 1988). The disequilibrium data of Williams et al. (1986) confirm that the natrocarbonatites are neither fused trona, nor the result of interaction between a magma and a groundwater reservoir of saline brine. A comparison of natrocarbonatites with evaporite from Lake Magadi showed marked differences in activities of certain isotopes from the U- and Th-decay series and U and Th abundances. Williams et al. (1986) argued that the differences in Th activity ratio preclude the possibility that the natrocarbonatite is remobilized trona-rich evaporite lake deposits. The $^{230}Th/^{232}Th$ activity ratio of the carbonatites is 1.00 ± 0.05, which represents equilibrium with material that has a $^{238}U/^{232}Th$ ratio of 1; the $^{230}Th/^{232}Th$ and $^{238}U/^{232}Th$ activity ratios of about 0.6 and 2.7 for a Lake Magadi evaporite are significantly different. In addition, the $^{234}U/^{238}U$ activity ratio of 1.50 is quite different from that of the natrocarbonatites, which is in secular equilibrium. Even if the natrocarbonatites could be considered as fused trona, the minimum time required to re-establish equilibrium in the U-series is about 0.75 million years, but in this time the activity ratio of $^{230}Th/^{232}Th$ would have reached 2.70, a value much greater than the value of 1 observed in the carbonatite. The disequilibrium studies rule out remobilization or fusion of trona, but not one involving dissolution of old trona followed by exsolution of natrocarbonatite. It will be apparent to those conversant with the history of carbonatite geology that the former creates a model which is strikingly similar to the one involving the remobilization of sedimentary limestones. Although the disequilibrium study certainly shows differences in some activity ratios between natrocarbonatite and trona, and this still demands an explanation, it does not rule out a model similar to the one proposed by Milton (1968), providing that the isotopic abundances of the silicate magma being contaminated remain relatively unaffected by the trona.

The enrichment of ^{228}Ra in natrocarbonatite argues against the hypothesis of Eugster (1970) involving interaction of alkaline magma with saline brines. Williams et al. (1986) maintain that there would be insufficient Th in the sediments to maintain sufficient amounts of ^{228}Ra, particularly in a convecting hydrothermal system. For these reasons, partial melting of the upper mantle or exsolution from a mantle-derived silicate magma were evaluated by Williams et al. (1986), whereas fusion or assimilation of near-surface sediments were not.

3 Conclusions

Like Baker (1958), we favour the theory that the sodium salts in Lakes Magadi and Natron were derived, in part, from leaching of the surrounding volcanic rocks, and this would include a contribution from the natrocarbonatites of Oldoinyo Lengai. During the 1966 eruption of the volcano, ash from Oldoinyo Lengai was showered over much of the Magadi and Natron drainage basins to the west, north-west and north of the volcano (Dawson et al. 1968). When this highly soluble ash is weathered, the potassium will be absorbed on zeolites (which are abundant in the tuffs covering the Salei and Serengeti Plains), whilst the sodium will be carried in solution to the soda lakes and concentrated there. The problem here is similar to the "chicken and egg" argument, as to whether natrocarbonatite is derived from the evaporites or vice versa.

We accept the fact that the addition of sodium-rich evaporites is one way of raising the sodium content of a sodium-poor magma, but the already reasonably high sodium content of the nephelinite flows, in particular combeite-bearing nephelinite BD 70 (13.5 wt% Na_2O, Donaldson et al. 1987) and others, means that an additional source of Na, such as evaporites, is not really needed, unless of course the combeite nephelinites themselves are the products of a contaminated magma. Extreme differentiation, such as the one proposed by Peterson and Kjarsgaard (Chap. 11, this Vol.), is supported by both experimental and observational data, although admittedly the Lengai natrocarbonatite is the result of an extreme process affecting a somewhat unusual magma type. Other problems that cast doubt on the "evaporite – saline brine" hypotheses involve the stable isotope composition of the trona, and the differences in K/Na ratios between trona and natrocarbonatite. The similarity of many trace elements from natrocarbonatites with those from known plutonic carbonatites, deficient in alkalis, argues for a carbonatite lineage for the Lengai flows.

On balance, we feel that there is more evidence, at present, to support the derivation of the natrocarbonatite from a magmatic source rather than a model involving continental evaporites. There are some features, admittedly, of the magmatic origin for natrocarbonatite that do present problems. The absence of natrocarbonatites at other nephelinitic centres is a point of concern, but this may reflect the rather extreme conditions needed for the development of a natrocarbonatite magma, or simply result from rapid solution of natrocarbonatite during the weathering cycle.

Acknowledgements. Both writers would like to acknowledge amicable correspondence and exchange of ideas with the late C. Milton over the years. His voicing of the "evaporite" hypothesis at the 1989 IGC in Washington forced us to reconsider his model and to give it, along with Eugster's, this long overdue public airing. Thanks are extended to J.W. Card, A. Simonetti, D.M. Pyle, and A.R. Woolley for their helpful comments during the preparation of this manuscript. Comments by the two reviewers, D.D. Hogarth and J. Wolff, and series editor Gail Mahood, are much appreciated.

References

Baker BH (1958) Geology of the Magadi area. Report 42, Geol Surv Kenya, 81pp
Barker DS (1989) Field relations of carbonatites. In: Bell K (ed) Carbonatites – genesis and evolution. Unwin Hyman, London, pp 38–69
Bell K, Blenkinsop J (1987) Nd and Sr isotopic compositions of East African carbonatites: implications for mantle heterogeneity. Geology 15:99–102
Bell K, Farquhar RM, Dawson JB (1973) Strontium isotope studies of alkalic rocks: the active carbonatite volcano Oldoinyo Lengai, Tanzania. Bull Geol Soc Am 84:1019–1030
Bottinga Y (1968) Calculation of fractionation factors for carbon and oxygen isotope change in the system calcite-carbon dioxide-water. J Phys Chem 72:800–808
Condomines M, Hemond Ch, Allègre CJ (1988) U-Th-Ra radioactive disequilibria and magmatic processes. Earth Planet Sci Lett 90:243–262
Dawson JB (1962) The geology of Oldoinyo Lengai. Bull Volcan 24:349–387
Dawson JB (1964) Reactivity of the cations in carbonate magmas. Proc Geol Soc Can 15:103–113
Dawson JB (1989) Sodium carbonate lavas from Oldoinyo Lengai, Tanzania: implications for carbonatite complex genesis. In: Bell K (ed) Carbonatites – genesis and evolution. Unwin Hyman, London, pp 255–277
Dawson JB (1993) A supposed sövite from Oldoinyo Lengai, Tanzania: result of extreme alteration of alkali carbonatite lava. Min Mag 57:93–101
Dawson JB, Powell V (1969) The Natron-Engaruku explosion crater area, northern Tanzania. Bull Volcan 33:791–817
Dawson JB, Pinkerton H, Norton GE, Pyle DM, Jackson D, Fallick A (1990) Alkali carbonatite lavas of 1988 eruption, Oldoinyo Lengai: petrology and geochemistry. IAVCEI volume, Mainz
Denaeyer ME (1970) Rapports isotopiques ^{8}O et ^{8}C et conditions d'affleurement des carbonatites de l'Afrique centrale. C R Acad Sci 270D:2155–2158
Donaldson CH, Dawson JB, Kanaris-Sotiriou R, Batchelor RA, Walsh NJ (1987) Thesilicate lavas of Oldoinyo Lengai, Tanzania. Neues Jahrb Miner Abh 156:247–279
Eugster HP (1970) Chemistry and origin of the brines of Lake Magadi, Kenya. Spec Pap Mineral Soc Am 3:215–235
Eugster HP (1980) Lake Magadi, Kenya and its precursors. In: Nissenbaum A (ed) Hypersaline brines and evaporites. Elsevier, Amsterdam, pp 195–232
Eugster HP (1986) Lake Magadi, Kenya: a model for rift valley hydrochemistry and sedimentation. In: Frostick LE et al. (eds) Sedimentation in the African Rifts. Geol Soc Spec Publ 25:177–189
Gittins J (1989) The origin and evolution of carbonatite magmas. In: Bell K (ed) Carbonatites – genesis and evolution. Unwin Hyman, London, pp 580–600
Gittins J (1991) The role of fluorine in carbonatite magma evolution. Nature 349:56–58
Guest NJ, Stevens JA (1951) Lake Natron, its springs, rivers, brines and visible saline reserves. Geol Surv Tanganyika Mineral Resour Pamphlet No 28
Hay RL (1966) Zeolites and zeolitic reactions in sedimentary rocks. Geol Soc Am Spec Pap 85:130
Isaac G LI (1965) The stratigraphy of the Peninj beds and the provenance of the Natron australopithicine mandible. Quaternaria 7:101–130
Javoy M, Pineau F, Staudacher T, Chaminée JL, Krafft M (1989) Mantle volatiles sampled from a continental rift: the 1988 eruption of Oldoinyo Lengai. TERRA Abstr 1:324
Keller J, Krafft M (1990) Effusive natrocarbonatite activity of Oldoinyo Lengai, June 1988. Bull Volcanol 52:629–645
Kjarsgaard BA, Hamilton DL (1989) The genesis of carbonatites by immiscibility. In: Bell K (ed) Carbonatites – genesis and evolution. Unwin Hyman, London, pp 388–404
Le Bas MJ (1989) Diversification of carbonatite. In: Bell K (ed) Carbonatites – genesis and evolution. Unwin Hyman, London, pp 428–447
Mariano AN, Roeder PL (1983) Kerimasi: a neglected carbonatite volcano. J Geol 91:449–455

Milton, C (1968) The "Natro-Carbonatite Lava" of Oldoinyo Lengai, Tanzania. Geol Soc. Am Program with Abstr, 202

Milton C (1989) Oldoinyo Lengai natrocarbonatite lava: its history. 28th Int Geological Congress, Washington 1989, pp 2–441

Niggli P (1916) Gleichgewichte zwischen TiO_2 und CO_2, sowie SiO_2 und CO_2 in Alkali-Kalk-Alkali, und Alkali-Aluminatschmelzen. Z Anorg Allg Chem 98:241–326

O'Neil JR, Hay RL (1973) $^{18}O/^{16}O$ ratios in cherts associated with the saline lakes of East Africa. Earth Planet Sci Lett 19:257–266

Peterson TD (1989) Peralkaline nephelinites I. Comparative petrology of Shombole and Oldoinyo Lengai, East Africa. Contrib Mineral Petrol 101:458–478

Peterson TD (1990) Petrology and genesis of natrocarbonatite. Contrib Mineral Petrol 105: 143–155

Peterson TD, Marsh BD (1986) Sodium metasomatism and mineral stabilities in alkaline ultramafic rocks: implications for the origin of the sodic lavas of Oldoinyo Lengai, EOS 67-16

Sheppard SMF, Dawson JB (1973) $^{13}C/^{12}C$ and D/H isotope variations in "primary igneous carbonatite". Fortschr Mineral 50:128–129

Suwa K, Oana S, Wada H, Osaski S (1975) Isotope geochemistry and petrology of African carbonatites. Phys Chem Earth 9:735–745

Vinogradov AP, Kropotova OI, Gerasimovsky VI (1970) Carbon isotope composition from carbonatites from East Africa. Geokhimiya 1970:643–646

Williams RW, Gill JB, Bruland KW (1986) Ra-Th disequilibrium systematics: timescale of carbonatite magma formation at Oldoinyo Lengai volcano, Tanzania. Geochim Cosmochim Acta 50:1249–1259

Woolley AR (1969) Some aspects of fenitization with particular reference to Chilwa Island and Kangankunde, Malawi. Bull Br Mus (Nat Hist) 2; 4:219

Woolley AR, Kempe DRC (1989) Carbonatites: Nomenclature, average chemical compositions, and element distribution. In: Bell K (ed) Carbonatites – genesis and evolution. Unwin Hyman, London, pp 1–14

What Are the Parental Magmas at Oldoinyo Lengai?

T.D. Peterson and B.A. Kjarsgaard

Abstract

At Oldoinyo Lengai, all of the carbonate lavas, and most of the silicate lavas, are highly fractionated with low Mg number (Mg#) and very high alkali content. Experimental studies have demonstrated that natrocarbonatite can be produced by liquid immiscibility from strongly peralkaline nephelinites, which are found at Oldoinyo Lengai in close association with natro-carbonatite. If natrocarbonatite is formed by liquid immiscibility from a CO_2-saturated silicate magma, the origin of that magma is clearly the outstanding petrogenetic problem at Oldoinyo Lengai. Existing 1-atm experimental data on the crystallization of primary alkaline, ultrabasic magmas (olivine nephelinite and olivine melilitite) and petrological studies on eruptive examples indicate that the derivatives of olivine melilitites and melilite nephelinites can be strongly peralkaline ($[Na+K]/Al \geqslant 1.5$) and frequently contain wollastonite, a common phenocryst phase in nephelinites at Oldoinyo Lengai. Olivine melilitites occur at Oldoinyo Lengai and at young tuff cones nearby; these have alkali contents that are abnormally high, even for these rocks. Olivine melilitites cannot fractionate to form phonolites; relatively low-alkali phonolites at Oldoinyo Lengai were probably formed by crystal fractionation and crustal contamination of olivine nephelinite, which is also sparsely represented there. Oldoinyo Lengai is an excellent example of extreme igneous fractionation, but cannot stand as a general model for carbonatite petrogenesis because its parental magmas are rare in alkaline centres.

1 Introduction

It is generally agreed that most of the eruptive rocks at the nephelinite-carbonatite volcano Oldoinyo Lengai represent highly fractionated magmas, since they are rich in alkalies and low in MgO. This is especially true for the Recent natrocarbonatite lavas and ashes, and silicate lava flows, lava blocks and lapilli tuffs (Donaldson et al. 1987; Twyman and Gittins 1987; Peterson 1989a). The natrocarbonatites are notable for having the highest $(Na+K)/(Na+K+Ca)$ ratio of any known carbonatite (average 0.7: Dawson 1962a). The most evolved silicate lavas at Oldoinyo Lengai, with $(Na+K)/Al > 7$ in the glass (Peterson 1989a), are the most peralkaline known. Any successful petrogenetic model for Oldoinyo Lengai must account for the exceptionally alkaline composition of most of its lavas, as well as the presence of two magma types: one silicate and one carbonate.

The Geological Survey of Canada, 588 Booth Street, Ottawa, Ontario, Canada K1A 0E4

It was earlier thought that natrocarbonatite might be a primary magma type (Dawson 1962b; Le Bas 1987). Experiments on the peridotite-C-O-H solidus at mantle pressures (Sweeney et al., Chap. 13, this Vol.) indicate that carbonate liquids produced under these conditions are magnesian and only moderately alkaline. Highly alkaline carbonate liquid can form under mantle P-T conditions but only from high-Na, low-Mg source rocks, which have not been observed as mantle nodules (Baker and Wyllie 1988). Therefore, if natrocarbonatite is ever primary, it is in the sense of an initial carbonatite melt derived by liquid immiscibility, or by partial melting of intrusive rocks. It is not possible to derive other carbonatite melts, such as sövite magma with calcite phenocrysts, from natrocarbonatite, since natrocarbonatite has much lower crystallization temperatures than other carbonatites (Cooper et al. 1975). However, subsolidus leaching of Na+K from a natrocarbonatite might produce *solid* sövite (Dawson et al. 1987). Most petrologists would agree that the natrocarbonatite of Oldoinyo Lengai was derived either by crystal fractionation from less alkaline carbonatite melt (e.g. Twyman and Gittins 1987; Wallace and Green 1988) or by liquid immiscibility from a fractionated silicate melt (Koster van Groos and Wyllie 1966; Freestone and Hamilton 1980).

In this chapter, we point out that the exceptionally alkaline character of Oldoinyo Lengai is mirrored in young carbonate-rich olivine melilitite tuff cones of the area (Dawson et al. 1985). Crystallization of olivine melilitites may be modelled in portions of the join Na-Ab-Ln-Fo (e.g. Pan and Longhi 1989, 1990). The relevent subsystems show a strong tendency towards peralkaline residual liquids (Yoder 1979), suggesting a possible connection between the tuff cones and the Recent alkaline lavas. Some of the strongly peralkaline silicate lavas of Oldoinyo Lengai correspond exactly to those compositions that can coexist with immiscible natrocarbonatite, as determined by numerous experiments on natural and synthetic compositions. This is consistent with the natrocarbonatite magma originating by exsolution from a highly evolved, CO_2-saturated nephelinite magma.

2 Summary Description of Oldoinyo Lengai Lavas

The silicate lavas from Oldoinyo Lengai have been described by Donaldson et al. (1987) and compared to those of Shombole, a late Pliocene nephelinite-carbonatite volcano 75 km north of Oldoinyo Lengai (Fig. 1) by Peterson (1989a). The natrocarbonatite lavas, consisting mainly of phenocrysts of the Na-K-Ca carbonates nyerereite and gregoryite in a matrix of gregoryite and halide minerals, have been described by Dawson (1962a,b), Peterson (1990), and Keller and Krafft (1990).

Representative analyses of silicate lavas from Oldoinyo Lengai are given in Table 1. Few samples of mafic lavas that might serve as models for parental magmas have been described; we have located only one analysis with Mg# > 0.6 (where Mg# = Mg/[Mg+Fe$_{total}$]) (Guest 1953, in Dawson 1966). This single high-Mg# lava (JG1279) is a melanephelinite with high CaO (14.4%); its

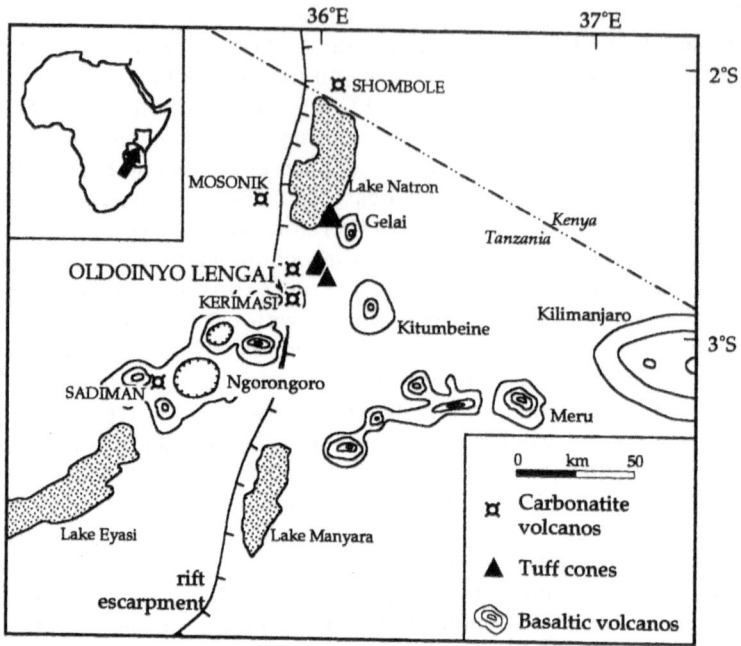

Fig. 1. Location of Oldoinyo Lengai and other volcanic centres of the south Kenya Rift. Recent olivine melilitite centres (tuff cones), from north to south, are Armykon Hill, Lalarasi and Oldoinyo Loolmurwak

mineralogy is unknown. There are only four other lavas recorded from Oldoinyo Lengai with Mg# ⩾ 0.25. Thus, Oldoinyo Lengai has mainly erupted highly fractionated rocks, and the nature of the parental magmas must be inferred from these.

According to Dawson (1962b, 1966), the earliest, most voluminous phase of activity at Oldoinyo Lengai produced nephelinites and phonolites (the yellow pyroclastics). The yellow pyroclastics are overlain by a thin veneer of black pyroclastic rocks, followed by minor melanephelinite flows. The black pyroclastics contain ijolitic rocks rich in wollastonite, interpreted to be mainly metasomatic in origin (Dawson 1966). Recent activity at Oldoinyo Lengai has been dominated by explosive eruptions and lava flows of natrocarbonatite (Nyamweru 1988). During the Recent phase of Oldoinyo Lengai, nephelinites have been erupted as melt lapilli and lava blocks with natrocarbonatite ash (Dawson 1962a,b), and some nephelinite flows exposed in the active crater wall are interbedded with natrocarbonatite flows (Keller and Kraft 1990). The nephelinites have high (Na+K)/Al (up to 2.3 for some whole rocks), with correspondingly unusual mineral assemblages. In addition to nepheline, clinopyroxene, titanite and melanite, they may contain abundant phenocrysts of sodalite and wollastonite or combeite ($Na_2Ca_2Si_3O_9$), as well as cancrinite, leucite, zeolites and primary carbonate (Donaldson et al. 1987; Peterson 1989a).

Table 1. Representative analyses of silicate lavas from Oldoinyo Lengai. MG = Mg/(Mg + Fe^{2+}) in normative ferromagnesian minerals. Data sources: Dawson et al. (1985), Dawson (pers. comm.), Donaldson et al. (1987), Peterson (1989b)

Rock	[a]Tuff cones JG-1279		BD 66	BD 81	HOL-16[b]	HOL-10[c]	BD 74
	Olivine melilitite	Melilite nephelinite	Wollastonite nephelinite	Mel-woll nephelinite	Wollastonite nephelinite	Combeite nephelinite	Phonolite
SiO_2	35.36	43.93	38.25	44.74	44.00	43.80	52.92
TiO_2	4.71	2.76	1.70	1.88	1.25	1.00	0.90
Al_2O_3	8.17	13.84	14.20	14.63	16.00	13.40	19.83
Fe_2O_3	8.75	2.40	11.10	7.21	5.60	5.00	3.66
FeO	4.73	5.42	2.07	1.88	2.60	4.30	1.78
MnO	0.23	0.44	0.29	0.23	0.24	0.36	0.17
MgO	9.58	7.47	2.25	2.02	0.99	0.63	0.66
CaO	16.56	14.43	13.04	8.70	7.15	6.97	2.88
Na_2O	4.30	5.59	7.74	8.06	10.90	14.90	10.21
K_2O	2.33	2.13	4.11	4.31	5.08	5.66	4.65
P_2O_5	1.64	0.51	0.59	0.63	0.35	0.33	0.18
CO_2	1.43	NA	2.41	1.29	2.20	0.80	0.09
H_2O	1.49	NA	2.28	3.52	3.30	1.40	1.48
Sum	99.26	98.92	99.85	99.10	99.66	98.55	99.41
(Na+K)/Al	1.20	0.83	1.23	1.23	1.46	2.29	1.10
Mg/(Mg+Fe)	0.55	0.64	0.25	0.30	0.19	0.11	0.23

Table 1. *Continued*

Rock	[a]Tuff cones JG-1279		BD 66	BD 81	HOL-16[b]	HOL-10[c]	BD 74
	Olivine melilitite	Melilite nephelinite	Wollastonite nephelinite	Mel-woll nephelinite	Wollastonite nephelinite	Combeite nephelinite	Phonolite
Anor	–	–	–	–	–	–	–
Orth	–	6.38	2.48	23.28	17.79	5.11	27.48
Albite	–	3.11	–	–	–	–	17.54
Kals	7.82	5.39	12.39	1.25	6.95	16.10	–
Neph	15.74	25.63	26.67	27.77	29.27	20.27	31.73
Na-met	–	–	–	–	–	16.68	–
Woll	7.15	–	7.61	1.61	4.58	0.61	–
Lrnt	6.46	–	–	–	–	–	–
Acmt	–	–	14.33	14.92	16.20	14.47	9.06
Diop	30.60	48.22	17.37	16.69	12.66	19.49	7.96
Oliv	6.78	0.38	–	–	0.94	–	0.57
Titan	–	–	–	–	–	–	2.21
Prvsk	2.35	0.68	2.89	3.20	2.13	1.70	–
Mgnt	4.90	3.48	2.69	1.36	–	–	0.77
Hemt	–	–	4.29	1.11	–	–	–
Ilmt	8.95	4.48	–	–	–	–	–
Aptit	3.80	1.18	1.37	1.46	0.81	0.76	0.42
Calcit	3.25	–	5.48	2.93	5.00	1.82	0.20
Sum	97.80	98.93	97.58	95.59	96.33	97.02	97.94
MG	1.00	0.83	0.72	0.68	0.38	0.19	0.41

[a] Average of three analyses. [b] Includes 0.20% F and 0.23% Cl. [c] Includes 0.42% F and 0.53% Cl.

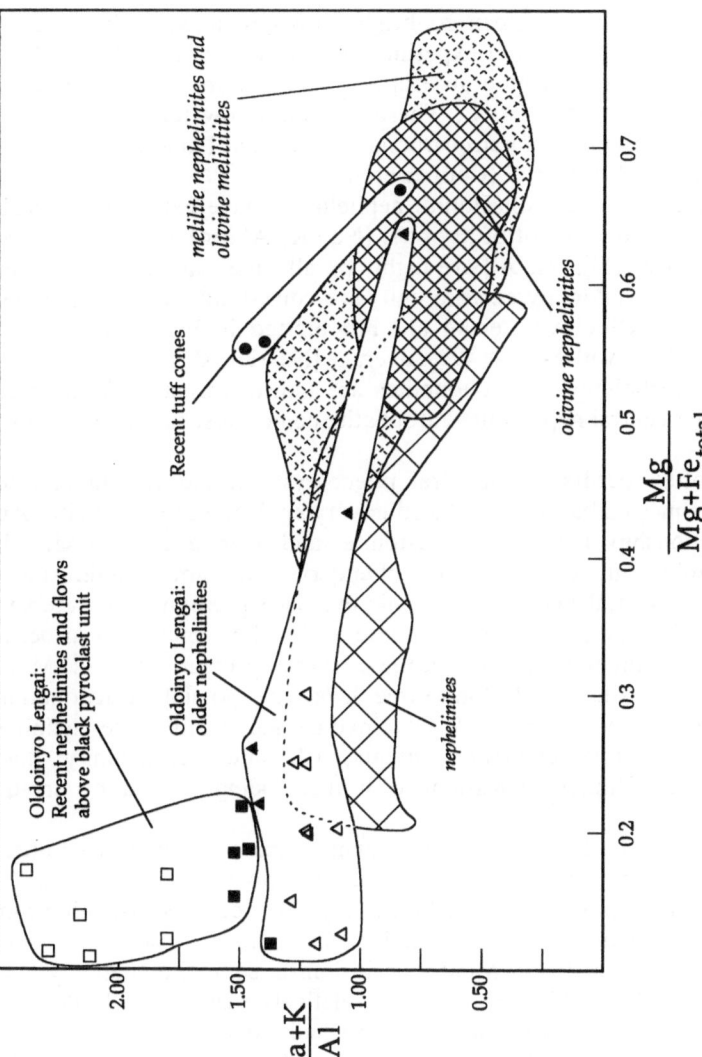

Fig. 2. Peralkalinity versus Mg/(Mg+Fe$_{total}$) (=Mg) for olivine nephelinites, olivine melilitites, and derivative lavas. *Open squares* Combeite nephelinites; *filled squares* wollastonite nephelinites; *open triangles* nephelinites; *filled triangles* melanephelinites; *filled circles* olivine melilitites. Melilite nephelinites and olivine melilitites overlap and have been combined into one field for clarity, however, melilite nephelinites are concentrated near the low-Mg end. The fields of Oldoinyo Lengai lavas and nearby Recent olivine melilitites (no patterns, with data points) includes all analyses (excluding phonolites) of Dawson (1966), Dawson et al. (1985), Donaldson et al. (1987), Peterson (1989a), and Keller and Krafft (1990). Data for other fields (*italicized labels*) is from King and Sutherland (1966), Wimmenauer (1966), Strong (1972), Bailey (1974), Le Bas (1977, 1987), Tazieff (1977), McIver (1981), Clague and Frey (1982), Mitchell and Platt (1983), Wilkinson and Stolz (1983), Donaldson et al. (1987), Francis and Ludden (1989), Peterson (1989a), Bednarz and Schmincke (1990), and Keller and Krafft (1990)

The glasses, which are brown-green to bright grass-green when fresh, contain high Cl (0.6%). Crystallization experiments and nepheline-liquid geothermometry (Peterson 1989a) indicated an eruption temperature for the combeite nephelinites in the range 550–750°C. These unusually low temperatures are consistent with the co-eruption of natrocarbonatite, which at 1 kbar has a liquidus temperature (nyerereite) of 650°C (Cooper et al. 1975).

The Recent combeite and wollastonite nephelinites have extreme compositions, as is demonstrated in a plot of Mg# vs. (Na+K)/Al for lavas from several alkaline (sodic) centres (Fig. 2). All are highly peralkaline; samples from older units (particularly the yellow pyroclastic unit) are consistently less peralkaline. The phonolites from Oldoinyo Lengai are not plotted in Fig. 2 for clarity; however, for all analyses with $SiO_2 \geqslant 49\%$, average [Na+K]/Al is 1.09 (maximum 1.15). The older nephelinites and phonolites are indistinguishable from those found at most other central nephelinite-carbonatite complexes, such as Shombole (Peterson 1989a).

Wollastonite often occurs as euhedral phenocrysts in the Recent silicate lavas, but is sometimes embayed or replaced in varying degrees by combeite. An antipathetic relationship between wollastonite and combeite is observed, and lavas containing euhedral combeite phenocrysts do not contain stable wollastonite. Experimental and chemographic evidence presented by Peterson (1989b) indicated that the reaction wollastonite + liquid = combeite can occur as a result of equilibrium crystallization in these lavas at pressures near 0.1 MPa, and that combeite-bearing, wollastonite-free lavas are possible fractionation products of wollastonite nephelinites. At Oldoinyo Lengai, combeite nephelinites have higher incompatible element contents, Cl, and peralkalinity than the wollastonite nephelinites, consistent with their being related by crystal fractionation.

Evidence for Recent silicate-carbonate magma mixing at Oldoinyo Lengai occurs in the form of lapilli showing mineral-liquid reactions. Combeite, together with unusually Fe-rich and Al-poor melilites (Fig. 3), occur as disequilibrium reaction coronas surrounding pyroxene and wollastonite in mixed carbonate-silicate tuffs (Dawson et al. 1989). Melilite inclusions in combeite phenocrysts in nephelinite flows also have high Fe, and are relatively rich in Na (Keller and Krafft 1990). The nephelinites are not iron-enriched, so we speculate that the high Fe/Al ratio in the melilites is an effect of the extreme peralkalinity of the liquid coexisting with combeite. All melilite associated with combeite has higher Fe/Mg than the normal igneous melilite trend for Na-rich olivine melilitites, which has been reproduced experimentally (Fig. 3). Relatively Fe-poor melilites lying within this trend occur within a cumulate pyroxenite block from Oldoinyo Lengai (Donaldson and Dawson 1978).

Young (<1 Ma) tuff cones near Oldoinyo Lengai have consistently erupted carbonate-rich tuffs (Dawson 1964a,b). Carbonatite-melilitite volcanism has been widespread in northern Tanzania since about 3.7 Ma (Hay 1978). Dawson et al. (1985), in a comparative study of olivine melilitites from South and East Africa, noted that the Recent occurrences near Oldoinyo Lengai (at Lalarasi, Oldoinyo Loolmurwak, and Armykon Hill) had unusually high (Na+K)/Al (1.25, vs. 0.75 for average olivine melilitites). Glasses in the olivine melilitites

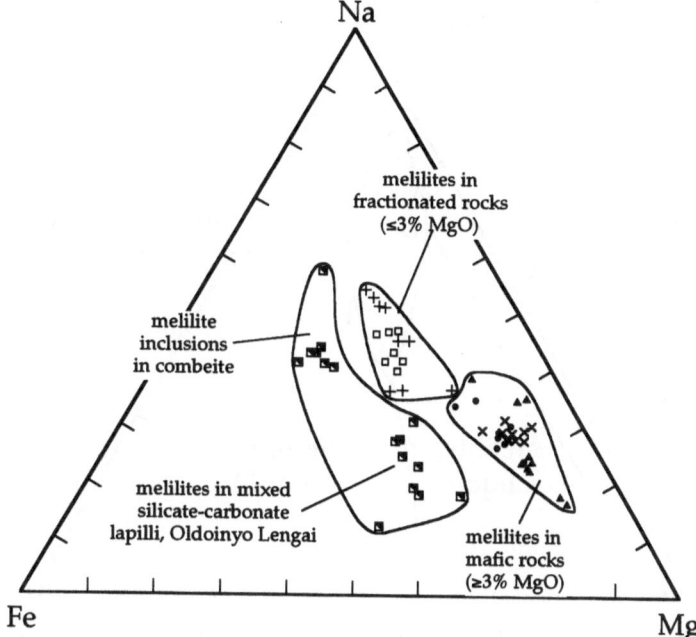

Fig. 3. Natural and experimental melilites from various alkaline rocks plotted in atomic Fe, Mg, and Na. *Open squares* Alkali pyroxenite nodule (Oldoinyo Lengai) (Donaldson and Dawson 1978); *half-filled squares* mixed silicate-carbonate lapilli from Oldoinyo Lengai (Dawson et al. 1989); *filled squares* inclusions in combeite phenocrysts (Keller and Krafft 1990); *triangles* olivine melilitites from Lalarasi, Oldoinyo Loolmurwak, and Armykon Hill (Dawson et al. 1985); *circles* miscellaneous olivine melilitites (Le Bas 1977; Deer et al. 1977; Hay 1978; McIver 1981; Wilkinson and Stolz 1983). Melilites from experimental charges (utilizing natural rocks as bulk compositions) with >3% MgO in the glass denoted by (×) and <3% MgO in the glass denoted by (+). (Data from Peterson 1987; Gee and Sack 1988; Kjarsgaard 1990)

are strongly peralkaline ([Na+K]/Al = 2.85) and have high volatile content, including 0.6% Cl (Dawson et al. 1985). It is reasonable to suggest that these flows may be closely related to the parental magmas of the recently erupted rocks at Oldoinyo Lengai, and this hypothesis can be tested by examining whether the generation of extremely peralkaline nephelinites, and their association with carbonate lavas with high (Na+K)/Ca, is consistent with an olivine melilitite parentage. This is the purpose of the following section.

3 Differentiation of Olivine Melilitites

Many alkaline magmas contain (F,OH)-bearing mafic phenocryst phases (mainly amphibole and phlogopite) and cannot be satisfactorily modelled in simplified, anhydrous systems. However, many nephelinites do not, including those from Oldoinyo Lengai and several other nephelinite-carbonatite centres in East

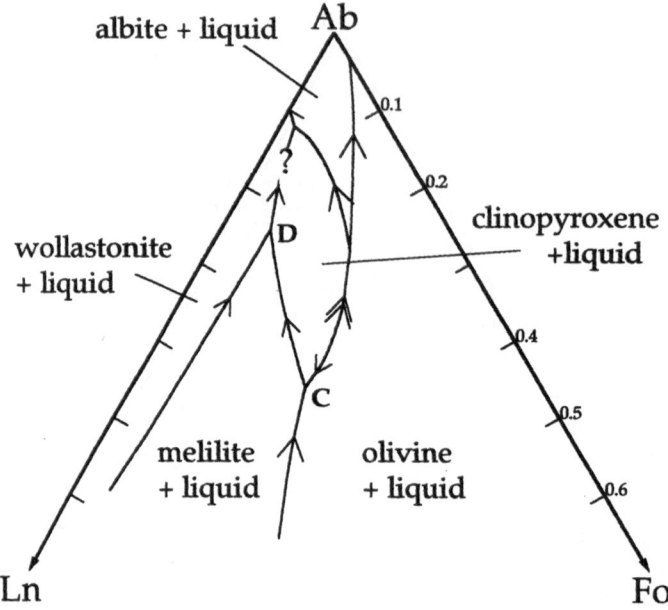

Fig. 4. Projection of liquidus surfaces in the system Ne-Ab-Ln-Fo from Ne, after Yoder (1979). Nepheline is an additional phase present in all liquidus fields. The *question mark* on the woll-cpx cotectic indicates that this cotectic is interpreted not to exist for strongly peralkaline compositions; the relevent cotectic should extend towards the liquidus field of combeite (which cannot be represented in this composition space). Note that the ol-cpx cotectic is a reaction curve for olivine, and contains a thermal divide

Africa, including Shombole (Peterson 1989b). Although (F,Cl,OH)-bearing phases such as sodalite, analcite, vishnevite and carbonate melt occur in highly fractionated rocks at these centres, the early crystallization of mafic nephelinitic magmas might be well approximated by volatile-absent systems.

The join Ne-Ab-Ln-Fo, a portion of the system SiO_2-Al_2O_3-MgO-CaO-Na_2O, contains compositions corresponding to olivine melilitite and olivine nephelinite magmas (Yoder 1979). Experimental work in this join and related ones (e.g. diopside-nepheline-sanidine: Platt and Edgar 1972) has established the relationship at low pressure between these two rock types, which are often found at the same volcanic centre (e.g. Wimmenauer 1966; Le Bas 1977). A crucial juncture is the pseudoinvariant point C (Fig. 4) where olivine (ol), melilite (mel), clinopyroxene (cpx) and nepheline (neph) are in equilibrium with liquid (liq). In liquidus surface projections from nepheline onto the plane Ab-Ln-Fo, three cotectics meet at this point, corresponding to the three magma types olivine melilite nephelinite (ol + mel + neph + liq), melilite nephelinite (mel + cpx + neph + liq), and olivine nephelinite (ol + cpx + neph + liq). Point C is pseudoinvariant because some solids that are present (e.g. aluminous clinopyroxene) are not contained within the join Ne-Ab-Ln-Fo.

Liquids on the ol-mel-neph cotectic move down-temperature toward point C, where olivine is in reaction relation wtih liquid (Yagi and Onuma 1978). If equilibrium is maintained and the reaction goes to completion, the magma then crystallizes with falling temperature along the mel-cpx-neph cotectic (melilite nephelinites). Unlike the other two cotectics, which meet at point C, the ol-cpx-neph cotectic contains a thermal divide; hence, crystallization paths both toward and away from point C are observed along it (Pan and Longhi 1989). However, the thermal divide lies very close to point C and therefore most olivine nephelinites take a crystallization path away from C, towards the pseudoinvariant point ol + cpx + neph + feldspar + liquid.

The ol + cpx + neph cotectic is a reaction curve for olivine; olivine has a reaction relation with liquid in olivine nephelinites at pressures to at least 27 kbar (Bultitude and Green 1971). The olivine can eventually disappear (aided by crystal fractionation) to produce nephelinites, which are relatively Mg-poor lavas containing phenocrysts of nepheline and clinopyroxene plus accessory phases such as titanite and melanite (Peterson 1989b). Continued crystallization will drive the nephelinite magma across the neph + cpx surface towards the liquidus field of alkali feldspar. Thus, olivine nephelinites can fractionate to cpx + neph (nephelinite) and then cpx + neph + san (phonolite) assemblages (Yoder 1979). Magmas richer in normative larnite than olivine nephelinite will follow a divergent crystallization path, distinguished by the precipitation of melilite.

It is thought that melilite nephelinites eventually attain a second reaction point (point D) where melilite reacts with liquid to produce wollastonite plus clinopyroxene (Platt and Edgar 1972), although to our knowledge, the reaction has not been demonstrated experimentally. Lavas recording this reaction have been elusive (see Yoder 1979) but Wilkinson and Stolz (1983) reported the presence of corroded melilite phenocrysts in melilite nephelinites from Oahu, and an increase in normative wollastonite. They also noted that residual liquids in these lavas (preserved as pegmatoids and irresolvable mineraloid patches) were markedly peralkaline, with $(Na+K)/Al$ up to 6. Although bulk compositions in the join Ne-Ab-Ln-Fo have $(Na+K)/Al=1$, all experimental studies have noted that residual liquids in this system tend to be peralkaline, due to crystallization of aluminous clinopyroxene (e.g. the system nepheline-diopside: Yoder and Kushiro 1972). Once peralkalinity is achieved, fractionation of nepheline, which has $(Na+K)/Al\approx1$, will increase it.

Experimental studies of the join Ne-Ab-Ln-Fo to date have not considered its relationship to the highly peralkaline residual liquids that it can generate. Such liquids cannot be properly described in that join, and would plot in Fig. 4 with a negative mass fraction of olivine due to the presence of normative acmite. Thus, the cotectic joining point D with the eutectic woll + neph + san + cpx + liq (wollastonite phonolites) probably does not exist for strongly peralkaline compositions. This interpretation is supported by the data of Wilkinson and Stolz (1983), who concluded that low-pressure differentiation of melilite-bearing rocks does not produce phonolitic compositions. Recent experimental data (Kjarsgaard et al., Chap. 12, this Vol.) indicate that, at pressures below 35 MPa, strongly peralkaline wollastonite nephelinite could fractionate to form combeite nephelinite. At higher pressures, and high CO_2 contents, Na_2O partitions into an immiscible alkali carbonate liquid before combeite saturation occurs. Thus, although combeite nephelinite and natrocarbonatite magmas may be closely associated, they are not conjugate liquids.

4 Discussion

The silicate lavas of Oldoinyo Lengai have a "Daly gap" in the Mg# (Fig. 2) and populations with high and low $(Na+K)/Al$ compositions occur on both sides of it. It is tempting to identify two differentiation trends in these lavas, distinguished by contrasting peralkalinity: one, originating in olivine nephelinite, produced the nephelinites and phonolites of the early eruptive phase; the other, starting with olivine melilitite, produced the wollastonite and combeite nephelinites of the Recent phase. Aside from Oldoinyo Lengai, combeite has only been recorded in lavas from Nyiragongo, Zaire (Sahama and Hytönen 1957), and has been documented from the Mayener Feld, Eifel, Germany (Fischer and Tillmans 1983). At both of these localities, olivine melilitites and/or melilite nephelinites are prominent primitive lavas (Sahama 1973; Tazieff 1977; Schmincke et al. 1983).

The model we have presented would have the natrocarbonatite of Oldoinyo Lengai originate by liquid immiscibility from strongly peralkaline wollastonite nephelinites (with combeite nephelinites as a low-pressure variant), which in turn were formed by crystal fractionation of parental olivine melilitite. We believe that all of the available data – geological, petrological, and experimental – is consistent with this model. Other conforming data could be cited; e.g. the radiogenic isotope compositions of natrocarbonatites of the 1988 eruption and combeite nephelinites are nearly identical (Keller and Krafft 1990). As a test of our model, evidence for highly alkaline carbonatites could be sought at other centres where melilite-bearing rocks are prominent (e.g. Oka, Québec; the Gardiner Complex, Greenland).

A plausible alternative to this model for the formation of the combeite nephelinites is that they were formed by mixing of very Na-rich material, such as trona or natrocarbonatite, with "normal" nephelinite magmas (Milton 1968; Peterson and Marsh 1986; Dawson et al. 1989). However, this model begs the question of the origin of natrocarbonatite and introduces circular arguments. We would predict that where natrocarbonatite is found, strongly peralkaline nephelinites will be close at hand. Other occurrences of "natrocarbonatite" have been described from Kerimasi, Tanzania; Kaluwe, Zambia; and Tinderet and Homa Mountain, Kenya (see Peterson 1990 for references and discussion). However, all those carbonatites are more calcic than natrocarbonatite and do not contain the phenocryst assemblage gregoryite + nyerereite (the interpreted assemblage being calcite + nyerereite). At present, the only known occurrence of true natrocarbonatite is at Oldoinyo Lengai.

Twyman and Gittins (1987), Gittins (1989) and Jago and Gittins (1990) have argued that natrocarbonatite is derived by crystal fractionation of less alkaline carbonatite magma, probably olivine sövite. Although not necessarily incon-sistent with our model, again, this assumption does not resolve the relationship between the silicate and carbonate magmas of Oldoinyo Lengai. In the system Na_2CO_3-K_2CO_3-$CaCO_3$ (Cooper et al. 1975), liquids in equilibrium with calcite (i.e. sövite magma) cannot fractionate to gregoryite-bearing assemblages because of a thermal divide at the composition of nyerereite$_{ss}$. This indicates that, in this simple ternary system, natrocarbonatite cannot be derived from sövite (Peterson 1990). Jago and Gittins (1990) have presented phase diagrams for the system Na_2CO_3-$CaCO_3$-CaF, in which the liquidus field of calcite expands across the composition of nyerereite. This permits a reaction relation between calcite and liquid to produce fluorite at the invariant point calcite + fluorite + nyerereite + liquid and, ultimately, the formation of the assemblage nyerereite + gregoryite + fluorite + liquid. While these phase relationships may be of relevance to extremely F-rich carbonatite magmas, we would argue that they are not applicable to Oldoinyo Lengai, because fluorite is not a phenocryst phase in natrocarbonatite. Fluorite does occur in the groundmass, but interpretations differ only as to whether it was the very last phase to crystallize (Keller and Krafft 1990) or the next to last (Peterson 1990). Fluorite also occurs within the immiscible sövite globules found in nephelinites at Shombole (Kjarsgaard and Peterson 1991) but only as a late-crystallizing phase. Fluorite is not present at all in the sövite lapilli of the Kaiserstuhl (Keller 1980) or in the fresh calcite carbonatite lavas at Fort Portal, Uganda (Barker and Nixon 1989), which are probably the most voluminous young carbonatite flows known next to those of Oldoinyo Lengai. There is therefore no evidence that early crystallization of fluorite constrained the magmatic evolution of carbonatites from Oldoinyo Lengai.

5 Summary

Oldoinyo Lengai has erupted silicate and carbonate magmas in two phases. During the early, most voluminous phase, nephelinites and phonolites of low peralkalinity [(Na+K)/Al≤1.3] were erupted, together with ijolite and syenite blocks from the underlying intrusive complex. In the Recent phase, strongly peralkaline nephelinites (but no phonolites) with rare mineral assemblages (wollastonite, combeite) were erupted with natrocarbonatite. Although olivine-bearing rocks, which might correspond to parental magmas, are rare or absent at Oldoinyo Lengai, unusually peralkaline olivine melilitites have erupted at several nearby small, young volcanic centers (e.g. Oldoinyo Loolmurwak). In the system Ne-Ab-Ln-Fo, liquids corresponding to olivine melilitites fractionate to peralkaline wollastonite nephelinites, suggesting that the primary magmas for the Recent phase of Oldoinyo Lengai were probably melilite-bearing. The older nephelinites and phonolites were probably formed by fractionation of olivine nephelinite magma. The CO_2-rich wollastonite nephelinites of Oldoinyo Lengai are of the correct composition to coexist with immiscible natrocarbonatite (Kjarsgaard et al., Chap. 12, this Vol.), and under most conditions would have exsolved natrocarbonatite during crystallization. At very low pressure, carbonatite would not have been exsolved and combeite phenocrysts would have eventually precipitated to form a distinctive rock (combeite nephelinite) closely associated with natrocarbonatite flows. Some combeite formed by disequilibrium reactions in mixed silicate-carbonate melt lapilli, but combeite nephelinites have mineral assemblages and textures consistent with near-equilibrium crystallization.

Hypotheses that describe natrocarbonatite as a crystal fractionation product of more Ca- or Mg-rich carbonatite magma fail because (1) the required bulk compositions and crystallization sequences do not resemble natural examples; (2) they do not resolve the relationship between natrocarbonatite and contemporaneous silicate magmas; and (3) less alkaline carbonate magmas have not been erupted at Oldoinyo Lengai, although they appear at many other centres where natrocarbonatite is not recorded. The Recent activity of Oldoinyo Lengai is an example of evolution of a relatively rare magma type (Na-rich olivine melilitite). Natrocarbonatite is not relevant to carbonatite genesis in general, because most nephelinite-carbonatite centres have olivine nephelinite as their parental magmas. The model developed in this chapter can be tested by looking for "runaway" peralkalinity trends and evidence of natrocarbonatite in centers where melilite-bearing rocks are prominent. One example may be the carbonate globules in häuyne-bearing melilite nephelinites of the Herchenberg volcano, East Eifel (Bednarz and Schminke 1990), which may prove to approach natrocarbonatite in composition.

Acknowledgements. T. Peterson gratefully acknowledges financial support from the Geological Society of America for field work in East Africa (1984–1986) while a Ph.D. student of Bruce Marsh at the Johns Hopkins University. Reviews by U. Mader, D. Green, and G. Brey improved the manuscript. This is GSC Contribution #13491.

References

Bailey DK (1974) Continental rifting and alkaline magmatism. In: Sørensen H (ed) The alkaline rocks. Wiley, London, pp 148–159
Baker MB, Wyllie PJ (1988) High pressure liquid immiscibility in the system nephelinite-$CaCO_3$-Na_2CO_3. Eos 69:1511
Barker DS, Nixon PH (1989) High-Ca, low-alkali carbonatite volcanism at Fort Portal, Uganda. Contrib Mineral Petrol 103:166–177
Bednarz U, Schmincke H (1990) Evolution of the Quaternary melilite-nephelinite Herchenberg volcano (East Eifel). Bull Volcanol 52:426–444
Bultitude RJ, Green DH (1971) Experimental study of crystal-liquid relationships at high pressures in olivine nephelinite and basanite compositions. J Petrol 12:121–147
Clague DA, Frey FA (1982) Petrology and trace element geochemistry of the Honolulu volcanics, Oahu: implications for the oceanic mantle below Hawaii. J Petrol 23:447–504
Cooper AF, Gittins J, Tuttle OF (1975) The system Na_2CO_3-K_2CO_3-$CaCO_3$ at 1 kilobar and its significance in carbonatite petrogenesis. Am J Sci 275:534–560
Dawson JB (1962a) Sodium carbonate lavas from Oldoinyo Lengai, Tanganyika. Nature 195:1075–1076
Dawson JB (1962b) The geology of Oldoinyo Lengai. Bull Volcanol 24:349–387
Dawson JB (1964a) Carbonate tuff cones in Northern Tanganyika. Geol Mag 101:129–137

Dawson JB (1964b) Carbonatitic volcanic ashes in Northern Tanganyika. Bull Volcanol 27:81–92

Dawson JB (1966) Oldoinyo Lengai – an active volcano with sodium carbonatite lava flows. In: Tuttle OF, Gittins J (eds) Carbonatites. Wiley, London, pp 155–168

Dawson JB, Smith JV, Jones AP (1985) A comparative study of bulk rock and mineral chemistry of olivine melilitites and associated rocks from East and South Africa. Neues Jahrb Min Abh 152:143–175

Dawson JB, Garson MS, Roberts B (1987) Altered former alkalic carbonatite lava from Oldoinyo Lengai, Tanzania. Inference for calcite carbonatite lavas. Geology 15:765–768

Dawson JB, Smith JV, Steele IM (1989) Combeite ($Na_{2.33}Ca_{1.74}others_{0.12}$) Si_3O_9 from Oldoinyo Lengai, Tanzania. J Geol 97:365–372

Deer WA, Howie RA, Zussman J (1977) An introduction to the rock-forming minerals, 10th edn. The Chaucer Press, Bungay, 528 pp

Donaldson CH, Dawson JB (1978) Skeletal crystallization and residual glass compositions in a cellular alkalic pyroxenite nodule from Oldoinyo Lengai. Contrib Mineral Petrol 67:139–149

Donaldson CH, Dawson JB, Kanaris-Sotiriou R, Batchelor RA, Walsh JN (1987) The silicate lavas of Oldoinyo Lengai. Neues Jahrb Min Abh 156:247–279

Fischer RX, Tillmans E (1983) Die Kristallstrukturen von natürlichem $Na_2Ca_2Si_3O_9$ vom Mt. Shaheru (Zaire) und aus dem Mayener Feld (Eifel). Neues Jahrb Min Abh 2:49–59

Francis D, Ludden J (1989) The mantle source for olivine nephelinite, basanite, and alkaline olivine basalt at Fort Selkirk, Yukon, Canada. J Petrol 31:371–400

Freestone IC, Hamilton DL (1980) The role of liquid immiscibility in the genesis of carbonatites: an experimental study. Contrib Mineral Petrol 73:105–117

Gee LL, Sack RO (1988) Experimental petrology of melilite nephelinites. J Petrol 29:1233–1255

Gittins J (1989) The origin and evolution of carbonatite magmas. In: Bell K (ed) Carbonatites – genesis and evolution. Unwin Hyman, London, pp 580–600

Hay RL (1978) Melilitite-carbonatite tuffs in the Laetolil beds of Tanzania. Contrib Mineral Petrol 67:357–367

Jago B, Gittins J (1990) Fluorine, carbonatite magma evolution and extrusive carbonatites. IAVCEI 1990 (Mainz) Abstr vol, 38 pp

Keller J (1980) Carbonatitic volcanism in the Kaiserstuhl alkaline complex: evidence for highly fluid carbonatitic melts at the Earth's surface. J Volcanol Geotherm Res 9:423–431

Keller J, Krafft M (1990) Effusive natrocarbonatite activity of Oldoinyo Lengai, June 1988. Bull Volcanol 52:629–645

King BC, Sutherland DS (1966) The carbonatite complexes of eastern Uganda. In: Tuttle OF, Gittins J (eds) Carbonatites. Wiley, London, pp 73–126

Kjarsgaard BA (1990) Nephelinite-carbonatite petrogenesis: experiments on liquid immiscibility in alkali silicate-carbonate systems. PhD Thesis, University of Manchester, Manchester

Kjarsgaard BA, Peterson TD (1991) Nephelinite-carbonatite liquid immiscibility at Shombole volcano, East Africa: petrographic and experimental evidence. Mineral Petrol 43:293–314

Koster van Groos AF, Wyllie PJ (1966) Liquid immiscibility in the system $Na_2O-Al_2O_3-SiO_2-CO_2$ at pressures up to 1 kilobar. Am J Sci 264:234–255

Le Bas MJ (1977) Carbonatite-nephelinite volcanism. Wiley, London, 347 pp

Le Bas MJ (1987) Nephelinites and carbonatites. In: Fitton JG, Upton BGJ (eds) Alkaline igneous rocks. Blackwell, London, pp 53–85

McIver JR (1981) Aspects of ultrabasic and basic volcanism intrusive rocks from Bitterfontein, South Africa. Contrib Mineral Petrol 78:1–11

Milton C (1968) The "Natro-Carbonatite Lava" of Oldoinyo Lengai, Tanzania. Geol Soc Am Program with Abstr, 202

Mitchell RH, Platt RG (1983) Primitive nephelinitic volcanism associated with rifting and uplift in the Canadian Arctic. Nature 303:609–612

Nyamweru C (1988) Activity of Oldoinyo Lengai, Tanzania, 1983–1987. J Afr Earth Sci 7:603–610

Pan V, Longhi J (1989) Low-pressure liquidus relations in the system $Mg_2SiO_4-Ca_2SiO_4-NaAlSiO_4-SiO_2$. Am J Sci 289:1–16

Pan V, Longhi J (1990) The system Mg_2SiO_4-Ca_2SiO_4-$CaAl_2O_4$-$NaAlSiO_4$-SiO_2: one atmosphere liquidus equilibria of analogs of alkaline mafic lavas. Contrib Mineral Petrol 105: 569–584

Peterson TD (1987) The petrogenesis and evolution of nephelinite-carbonatite magmas. PhD Thesis, The Johns Hopkins University, Baltimore

Peterson TD (1989a) Peralkaline nephelinites. I. Comparative petrology of Shombole and Oldoinyo Lengai, East Africa. Contrib Mineral Petrol 101:458–478

Peterson TD (1989b) Peralkaline nephelinites. II. Low pressure fractionation and the hypersodic lavas of Oldoinyo Lengai. Contrib Mineral Petrol 102:336–346

Peterson TD (1990) Petrology and genesis of natrocarbonatite. Contrib Mineral Petrol 105:143–155

Peterson TD, Marsh BD (1986) Sodium metasomatism and mineral stabilities in alkaline ultramafic rocks: implications for the origin of the sodic lavas of Oldoinyo Lengai. Eos 67:389–390

Platt RG, Edgar AD (1972) The system nepheline-diopside-sanidine and its significance to the genesis of melilite- and olivine-bearing alkaline rocks. J Geol 80:224–236

Sahama TG (1973) Evolution of the Nyiragongo magma. J Petrol 14:33–48

Sahama TG, Hytönen K (1957) Götzenite and combeite, two new minerals from the Belgian Congo. Min Mag 31:503–510

Schmincke H, Lorenz V, Seck HA (1983) The Quaternary Eifel volcanic fields. In: Fuchs K et al. (eds) Plateau uplifts. Springer, Berlin Heidelberg New York, pp 139–151

Strong DF (1972) Petrology of the island of Moheli, western Indian Ocean. Geol Soc Am Bull 83:389–406

Tazieff H (1977) An exceptional eruption: Mt. Nyiragongo, Jan. 10th, 1977. Bull Volcanol 40:1976–1977

Twyman JD, Gittins J (1987) Alkalic carbonatite magmas: parental or derivative? In: Fitton JG, Upton BGJ (eds) Alkaline igneous rocks. Blackwell, London, pp 85–94

Wallace ME, Green DH (1988) An experimental determination of primary carbonatite magma composition. Nature 335:343–346

Wilkinson JFG, Stolz AJ (1983) Low-pressure fractionation of strongly undersaturated alkaline ultrabasic magma: the olivine-melilite-nephelinite at Moiliili, Oahu, Hawaii. Contrib Mineral Petrol 83:363–374

Wimmenauer W (1966) The eruptive rocks and carbonatites of the Kaiserstuhl, Germany. In: Tuttle OF, Gittins J (eds) Carbonatites. Wiley, London, pp 183–204

Yagi K, Onuma K (1978) Genesis and differentiation of nephelinitic magma. Bull Volcanol 41:466–472

Yoder HS (1979) Melilite-bearing rocks and related lamprophyres. In: Yoder HS (ed) The evolution of the igneous rocks. Princeton University Press, Princeton, pp 391–412

Yoder HS, Kushiro I (1972) Composition of residual liquids in the nepheline-diopside system. Carnegie Inst Wash Year Book 71:413–416

Peralkaline Nephelinite/Carbonatite Liquid Immiscibility: Comparison of Phase Compositions in Experiments and Natural Lavas from Oldoinyo Lengai

B.A. Kjarsgaard[1], D.L. Hamilton[2], and T.D. Peterson[1]

Abstract

New low-pressure, low-temperature experiments (50–375 MPa, 700–850 °C) utilizing starting compositions of carbonated peralkaline nephelinite have generated immiscible carbonatite liquids that are alkali-rich. The results suggest the P-T conditions operating to produce the Oldoinyo Lengai natrocarbonatite lavas are quite low i.e. $\leq$100 MPa and $\leq$750 °C. These low temperatures are consistent with both calculated (silicate) and measured (carbonate) eruption temperatures for lavas from Oldoinyo Lengai. The conjugate silicate liquids are of peralkaline wollastonite nephelinite composition. The differentiation of peralkaline wollastonite nephelinite strongly depends on both pressure and to what degree the melt is saturated in CO_2. High pressure coupled with CO_2 saturation favours formation of an immiscible alkalic carbonate liquid, which becomes increasingly K_2O-rich with lower pressures of exsolution (P $\leq$ 200 MPa). Exsolution of an alkali-rich carbonate liquid from a silicate melt during further cooling and differentiation buffers the peralkalinity of the residual silicate liquid at near constant peralkalinity ([Na+K]/Al$\approx$2.15). At very low pressure (P $\leq$ 35 MPa) and CO_2 undersaturated conditions, carbonate liquid is not exsolved, and wollastonite nephelinite fractionate to exceptionally peralkaline [Na+K]/Al=4.0–7.0) combeite-bearing nephelinite. Liquid natrocarbonatite and combeite nephelinite are not conjugate, but may be expected to be closely associated. Alkalic carbonate liquids exsolved at low temperatures ($\leq$800 °C) are saturated with ferromagnesian solid phase(s) and are not superheated. Differentiation of alkalic carbonate melts (after these liquids have separated from their silicate host), consisting of fractionation of previously formed ferromagnesians, coupled with the precipitation of nyerereite+gregoryite produces a halogen-rich and SiO_2, TiO_2, Al_2O_3, MgO and FeO depleted natrocarbonatite magma.

1 Introduction

The origin of the natrocarbonatite lavas of Oldoinyo Lengai has been controversial and problematic since their discovery (Dawson 1962a). Koster van Groos and Wyllie (1963) suggested on the basis of preliminary results from their pioneering research on silicate/carbonate liquid immiscibility that the silicate and alkali carbonate lavas at Oldoinyo Lengai could be natural two-liquid analogues to the two liquids of their experiments. This theme was expanded on in a series of susbsequent publications (Koster van Groos and Wyllie 1966, 1968, 1973). Hamilton et al. (1979, 1989) and Freestone and Hamilton (1980) arrived

---

[1] Geological Survey of Canada, 601 Booth Street, Ottawa, Ontario, K1A 0E8, Canada
[2] Department of Geology, The University of Manchester, Manchester, M13 9PL, UK

at similar conclusions in their immiscibility studies, the first to use natural lavas from Oldoinyo Lengai as starting materials for two-liquid experiments.

The experiments reported in this chapter were conducted within the PT range thought, from other direct and inferred evidence, to have been in operation at the Oldoinyo Lengai volcano. The reported results support the arguments in favour of an origin by liquid immiscibility for the natrocarbonatite lavas. The new experiments, utilizing peralkaline wollastonite and combeite nephelinites (± synthetic natrocarbonatite) illustrate that liquid immiscibility can occur at low pressures (50–375 MPa) and low temperatures (700–850 °C). These P-T conditions are considerably lower than those used by Freestone and Hamilton (1980) and can be considered close to those at the Oldoinyo Lengai volcano. A genetic model for both the silicate and carbonate rocks of Oldoinyo Lengai is based on analyses of phenocrysts and quenched liquids from experimental charges and recently erupted lavas.

2 Experimental and Analytical Methods

2.1 Starting Materials

The majority of the experiments were based on starting materials using wollastonite nephelinite lava HOL14 (Peterson 1989a) ± synthetic natrocarbonatite. HOL14 is described in Peterson (1989a), but note that this sample also contains small (10–30 μm) combeite and sanidine crystals (see Fig. 1). A few additional experiments were based on combeite nephelinite HOL6 (Peterson 1989a) ± synthetic natrocarbonatite. Modal analyses of these two samples are listed in Table 1. There are many reasons for the use of highly peralkaline starting materials. Previous experimental work (Freestone and Hamilton 1980) had illustrated that only hyperalkaline silicate liquids had immiscible natrocarbonatite liquids. Peterson (1989a) noted that the occurrence of highly peralkaline nephelinites at volcanic centres other than Oldoinyo Lengai is rare, and suggested that the existence of natrocarbonatite is intimately associated with hyperperalkaline silicate lavas. Furthermore, these rocks contain a glassy groundmass that is exceptionally peralkaline. From an experimental view, this is a key objective in that the groundmass represents quenched silicate melt, and compre-

---

Fig. 1. *1* Back-scattered electron image of lava HOL14 used as a starting material. Phases are wollastonite (*Wo*), sodalite (*Sod*), nepheline (*Ne*), clinopyroxene (*Cpx*), combeite (*Co*), sanidine (*San*) and peralkaline silicate glass (*Ls*). *2* Back-scattered electron image of sealed-tube sample BK430 (P = 106 MPa, T = 750 °C, t = 290 h; bulk composition = HOL14*-10). Immiscible carbonate liquid (*Lc*) has quenched to a characteristic dendritic habit (large segregation at top of field of view and spherical globules in Ls). Solid phases are of nepheline, clinopyroxene, melanite garnet (*Mel*) and vishnevite (*Vi*) in peralkaline silicate glass. Phases labelled as in *1*. *3a* Backscattered electron image of an unpolished sample from sealed tube run BK431 (P = 280 MPa, T = 700 °C, t = 121 h; bulk composition = HOL14*-10), showing central

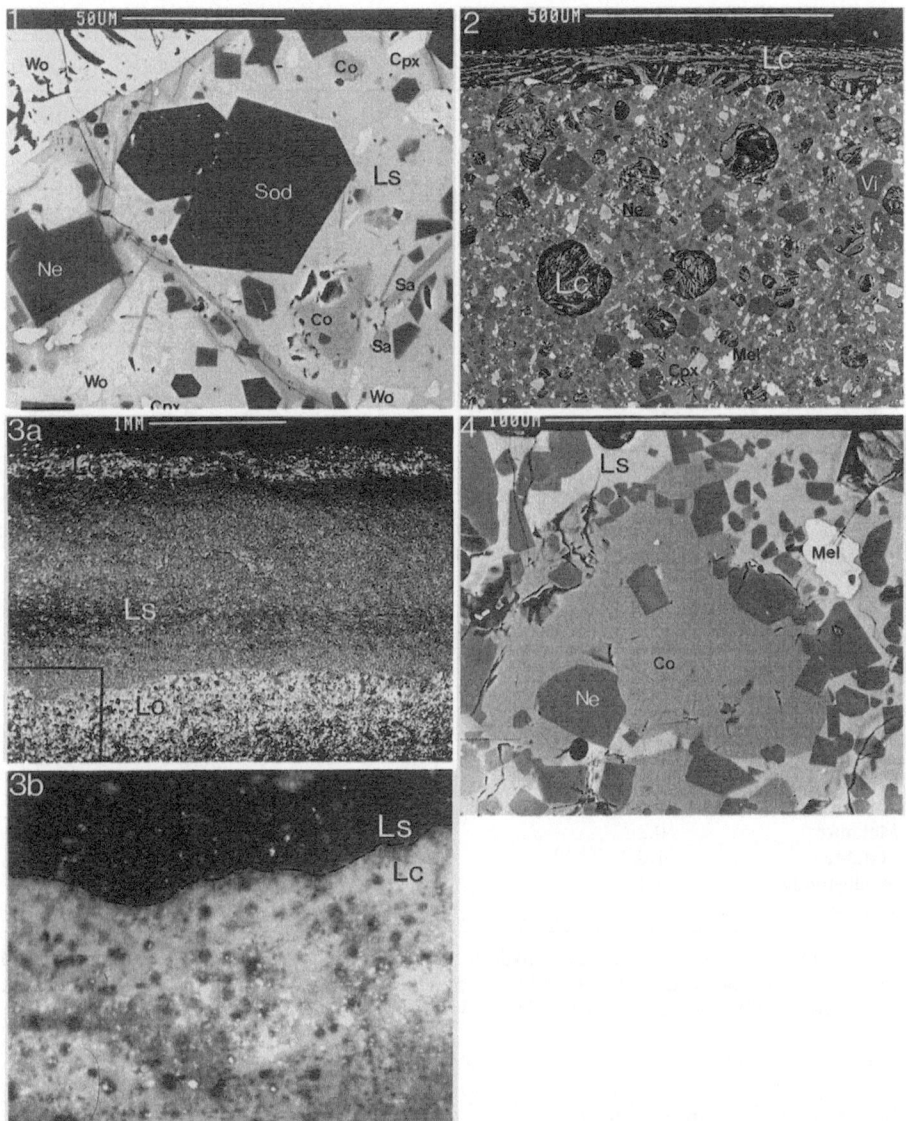

area of silicate liquid (+solids) with large segregations of immiscible carbonate liquid (+solids) at the edges. Phases labelled as in *1* and *2*. *3b* Photomicrograph of sample BK431 (enlargement of the area outlined in the *lower left corner* of *3a*). Euhedral crystals (*black*) in the immiscible carbonate liquid (*white/grey*; *bottom of photograph*) are of magnetite, melanite garnet and aegirine-augite. The meniscus between the immiscible silicate and carbonate liquids is shown by a *dashed black line*. Phases labelled as in *1* and *2*. *4* Back-scattered electron image of open tube run BK417 (P = 51 MPa, T = 700 °C, t = 290 h; bulk composition = HOL6). Solid phases are combeite (*Co*), nepheline (*Ne*) and melanite (*Mel*) garnet in hyper-peralkaline silicate glass.

Table 1. Bulk compositions and modal analysis of starting material lavas HOL6 and HOL14

	HOL14*	HOL14*-5	HOL14*-10	HOL6	HOL6-10
SiO_2	45.24	42.89	40.54	43.20	38.89
TiO_2	0.95	0.90	0.85	1.03	0.93
Al_2O_3	15.64	14.83	14.00	14.40	12.96
Fe_2O_3	5.54	5.25	4.96	6.20	5.61
FeO	3.17	3.02	2.87	3.40	3.06
MnO	0.29	0.28	0.28	0.36	0.35
MgO	0.90	0.87	0.83	0.70	0.65
CaO	5.98	6.36	6.73	6.55	7.28
Na_2O	11.19	11.86	12.52	11.70	13.56
K_2O	4.76	4.95	5.13	6.14	6.18
P_2O_5	0.28	0.31	0.34	0.40	0.47
H_2O	2.97	3.42	3.86	2.60	2.79
CO_2	1.49	3.07	4.65	2.10	5.28
F	0.22	0.36	0.46	0.38	0.46
Cl	0.27	0.40	0.58	0.41	0.61
S	0.38	0.43	0.47	0.76	0.80
SrO	0.34	0.39	0.45	0.41	0.51
BaO	0.27	0.30	0.34	0.38	0.44
Total	99.88	99.89	99.86	101.12	100.86
(Na+K)/Al	1.51	1.69	1.87	1.80	2.24

Modal analyses					
	HOL14			HOL6	
Groundmass	41.5			38.0	
Globules/Vesicles	19.4			12.8	
Nepheline	28.4			31.6	
Clinopyroxene	7.2			4.3	
Sphene	0.4			Trace	
Melanite	0.2			1.4	
Sodalite	1.8			2.0	
Wollastonite	1.4			0.4	
Combeite	Trace			9.5	

Note: HOL14 + 1 wt% oxalic acid dihydrate added = HOL14*; HOL14*-5 and HOL14*-10 have 1 wt% oxalic acid dihydrate added +5 or 10 wt% synthetic natrocarbonatite added, respectively; HOL6 + 10 wt% synthetic natrocarbonatite added = HOL6-10.

hension of the immiscibility problem requires examination of liquid compositions, not rock compositions.

The starting compositions are listed in Table 1 and consist of mixtures of HOL14 and oxalic acid ± synthetic natrocarbonatite (HOL14*, HOL14*-5, HOL14*-10) or HOL6 ± synthetic natrocarbonatite (HOL6 and HOL6-10). Starting materials HOL14*-5 and HOL14*-10, HOL6-10 contain 5 and 10 wt% added synthetic natrocarbonatite, respectively (made up of the following analytical grade reagents: $CaCO_3$, Na_2CO_3, K_2CO_3, NaF, KCl, $Ca_3(PO_4)_2$, $SrCO_3$ and $BaSO_4$). Lava HOL14 contains 3.40 wt% H_2O and 0.80 wt% CO_2 while lava HOL6 has 2.60 wt% H_2O and 2.40 wt% CO_2. The HOL14* series experiments have 1 wt% oxalic acid dihydrate added (analytical grade reagent) for two reasons. One was to increase the CO_2 level for the experiments without added natrocarbonatite, in an attempt to ensure saturation of the melt in this component. Additionally, there needs to be compensation for the fluid phase thought to be present with the magma at pressure, i.e. the lavas used as starting materials are vesicular and were probably in equilibrium with a fluid, but this latter is not "preserved" in the samples.

Mechanical mixtures of the starting materials were weighed out, hand ground in an agate mortar and redried before loading into gold tubes (3 mm OD × 20 mm length). Most experiments contained about 0.10 g of starting materials. Four of the experiments were run in unsealed (open) tubes. All experiments were performed at temperatures of 850 °C or less. Under these conditions, iron loss from the sample to the capsule is not a problem.

2.2 Experimental Method

Sample charges were run in either either internally heated pressure vessels (IHPV) for higher PT experiments or Tuttle-type (externally heated) pressure vessels for lower PT experiments at the University of Manchester. Argon was the pressure medium. Run times varied from 20 to 290 h. All samples were run unbuffered with respect to oxygen fugacity.

Details for IHPV experiments are the same as those described in Kjarsgaard and Hamilton (1989). For Tuttle vessel experiments, temperature was measured with a regularly calibrated single sheathed chromel-alumel thermocouple located in a well at the base of the vessel. Reported temperatures are believed accurate to ±10 °C. Pressure was measured using a calibrated Bourdon tube gauge, and is thought to be accurate to ±10 MPa. Samples were quenched at the end of the run by a jet of compressed air. Quenching rates were typically 300 °C/min.

2.3 Analysis

After the run, capsules were weighed to check for weight loss or gain during the run, punctured, and then re-weighed. Wieght loss after puncturing indicated that all sealed tube charges co-existed with a fluid phase. Charge contents were recovered in as large a piece as possible, to assist in textural interpretation. Run products were examined with a binocular microscope, and a small portion of the charge was used to make a grain mount for optical examination. Approximately half of the sample was used to make a polished thin section or set in resin to give a polished block for electron microprobe analysis. All samples were prepared in paraffin (kerosene) to prevent dissolution of alkali-bearing carbonates and/or salts.

Compositions of all solid and quenched liquid phases were determined using a Cameca Camebax electron microprobe (Geological Survey of Canada), using WDS techniques. Solid phases were analyzed with a slightly defocussed (5–10 µm) beam and specimen current of 10–30 nA. Quench silicate and carbonate liquid analyses utilized a tightly focussed (2–3 µm) beam which was rastered over an area (min = $100\,\mu m^2$; max = $2500\,\mu m^2$) and a specimen current of 10 nA. Counting times for all elements was 10 s. An accelerating potential of 15 kV was used for all analyses. Raw data were corrected with PAP software.

3 Results

Table 2 shows the results from the critical runs. No attempt was made to determine liquidus or solidus temperature of the bulk compositions, and T-X diagrams were not constructed due to lack of data points (see Kjarsgaard and Peterson (1991) for T-X diagrams in similar CO_2-rich nephelinite systems). The phase relationships are complex. In sealed-tube charges, run products consist of clinopyroxene + nepheline + melanite garnet + vishnevite ± wollastonite ± sanidine and one or two quenched liquid(s) with a fluid phase. The composition of the co-existing fluid phase was not determined, but on the basis of previous (unpublished) work, it is thought to be a mixed CO_2/H_2O fluid that is CO_2-rich. In open-tube experiments at 51 MPa and 700 °C, run products consist of clino-pyroxene + nepheline + melanite garnet ± vishnevite ± combeite ± wollastonite ± leucite ± sanidine and quenched silicate liquid (which is halogenrich). The assemblages from open-tube experiments closely resemble those described for the natural lavas (Peterson 1989a; this chapter). It is important to note that in runs with two liquids, solid phases are found in both quenched liquids (see Figs. 1, 3b), and that this is an important test as to whether the liquids are an immiscible pair. The exception to this is that nepheline, vishnevite and sanidine (all high-Al phases) are found only in the silicate liquid. Potential reasons for this are discussed in Kjarsgaard and Peterson (1991).

3.1 Solid Phases

Nepheline formed hexagonal prisms. The data is graphically illustrated in Fig. 2. Nepheline from HOL14-series sealed-tube experiments have a relatively restricted compositional range ($Ne_{70.8-74.0}$; $Ks_{18.2-20.7}$; $Qz_{0.5-4.5}$) over the pressures and temperatures of the study (Table 3, Fig. 2). These overlap with the range shown by unzoned phenocryts and microphenocrysts from wollastonite nephelinite HOL14 ($Ne_{71.2-73.3}$; $Ks_{20.0-20.4}$; $Qz_{2.7-3.3}$). In contrast, nepheline from HOL14- and HOL6-series open-tube 50 MPa experiments are much richer in kalsilite component ($Ne_{65.5-68.0}$; $Ks_{25.3-27.3}$; $Qz_{0.9-2.6}$) and overlap the compositional range shown by unzoned phenocrysts and microphenocrysts from combeite nephelinite HOL6 ($Ne_{64.4-68.8}$; $Ks_{23.2-24.7}$; $Qz_{1.3-2.8}$), although they are not quite as iron-rich.

Clinopyroxene formed stubby to acicular crystals. All pyroxenes are diopside – hedenbergite – aegirine solid solutions low in Ti and Al (Table 4). Clinoyroxene ferrous/ferric ratios were calculated utilizing the method of Droop (1987), and recast as ternary end-members (Fig. 3). Clinopyroxene from HOL14-series sealed-tube experiments are all classified as aegirine-augites (Morimoto 1988),

Table 2. Run data

Run no.	Bulk comp.	P (MPa)	T (°C)	T (h)	Phase assemblage
			Sealed-tube experiments		
BK398	HOL14*	375	850	38	LS + F
BK399	HOL14*-5	375	850	38	LS + F
BK400	HOL14*-10	375	850	38	LS + LC + F
BK411	HOL14*	380	750	50	LS + Cpx + Ne + Mela + Vish + F
BK412	HOL14*-5	380	750	50	LS + LC + Cpx + Ne + Mela + Vish + F
BK413	HOL14*-10	380	750	50	LS + LC + Cpx + Ne + Meal + F
BK425	HOL14*	280	700	121	LS + Cpx + Ne + Mela + Vish + F
BK428	HOL14*-5	280	700	121	LS + LC + Cpx + Ne + Mela + Vish + Mag + F
BK431	HOL14*-10	280	700	121	LS + LC + Cpx + Ne + Mela + Vish + Mag + F
BK424	HOL14*	106	750	290	LS + Cpx + Ne + Mela + Vish + F
BK427	HOL14*-5	106	750	290	LS + LC + Cpx + Ne + Mela + Vish + F
BK430	HOL14*-10	106	750	290	LS + Cpx + Ne + Mela + Wo + Vish + F
BK436	HOL14*	108	700	130	LS + LC + Cpx + Ne + Mela + Wo + Vish + San + F
BK429	HOL14*-5	108	700	130	LS + LC + Cpx + Ne + Mela + Wo + Vish + F
BK437	HOL14*-10	108	700	130	LS + LC + Cpx + Ne + Mela + Wo + Vish + F
			Open-tube experiments		
BK426	HOL14*	51	700	290	LS + Cpx + Ne + Mela + Wo + San
BK432	HOL14*-10	51	700	290	LS + Cpx + Ne + Mela + Vish + Comb
BK417	HOL6	51	700	290	LS + Cpx + Ne + Mela + Comb
BK423	HOL6-10	51	700	290	LS + Cpx + Ne + Mela + Leuc + Comb

Note: LS = silicate liquid; LC = carbonate liquid; F = fluid (CO_2/H_2O mixture; CO_2-rich); Cpx = clinopyroxene; Ne = nepheline; Mela = melanite garnet; Vish = vishnevite; Mag = magnetite; Wo = wollastonite; San = sanidine; Comb = combeite; Leuc = leucite.

Table 3. Representative microprobe analyses of nepheline from experimental charges and lavas HOL6 and HOL14

Sample	BK411	BK412	BK413	BK425	BK428	BK431	BK424	BK427	BK430	BK436	BK429	BK437	BK417	BK423	BK426	BK432	HOL6	HOL6	HOL6	HOL6	HOL6	HOL14	HOL14	HOL14
Type	ec	ec	ec	ec	ec	ec	ec	ec	ec	ec	ec	ec	ec	ec	ec	ec	mph	mph	mph	ph-c	ph-r	mph	ph-c	ph-r
SiO_2	41.93	43.17	43.83	42.50	42.56	43.17	42.43	41.62	41.57	42.66	42.23	42.86	41.84	42.12	41.20	40.95	41.27	41.38	41.7	41.82	41.99	42.68	42.67	42.58
Al_2O_3	33.50	33.26	31.82	32.42	32.5	32.21	32.75	33.62	33.20	32.39	31.99	32.96	31.51	32.31	32.61	32.59	31.06	32.35	30.81	32.73	32.66	32.24	33.30	32.75
Fe_2O_3	1.38	1.66	2.50	1.86	2.00	2.40	2.35	1.91	2.12	2.11	2.36	1.75	2.34	2.26	1.68	2.09	4.40	2.71	4.37	2.06	2.16	2.18	1.26	1.64
CaO	0.14	0.15	0.07	0.24	0.16	0.10	0.17	0.12	0.13	0.15	0.07	0.05	0.04	0.07	0.09	0.03	0.03	0.08	0.06	0.08	0.07	0.13	0.23	0.23
Na_2O	17.48	17.60	17.67	17.04	17.46	17.07	16.52	17.01	17.02	16.60	16.96	17.03	16.13	15.97	15.88	15.06	15.94	15.78	15.81	15.88	15.76	16.66	16.59	16.48
K_2O	6.74	6.10	6.01	5.88	6.43	5.93	6.24	6.63	6.54	6.70	6.50	6.23	8.09	8.24	8.27	8.70	7.86	7.85	7.78	7.95	7.87	6.50	6.51	6.59
Total	101.17	101.94	101.90	99.94	101.11	100.88	100.46	100.91	100.58	100.61	100.11	100.88	99.95	100.97	99.73	99.42	100.56	100.15	100.53	100.52	100.51	100.39	100.56	100.27
Ne	73.7	74.0	70.8	73.2	71.7	72.5	73.1	74.6	73.9	70.8	71.2	74.0	65.5	66.5	68.0	67.1	65.1	68.4	64.4	68.8	65.8	71.2	73.3	71.9
Ks	20.7	18.5	18.3	18.2	19.8	18.2	19.2	20.4	20.2	20.7	20.2	19.1	25.3	25.5	25.9	27.3	24.6	24.5	24.3	24.7	23.2	20.1	20.0	20.4
Nf	2.5	3.0	4.5	3.4	3.6	4.3	4.3	3.5	3.9	3.8	4.3	3.2	4.8	4.6	3.5	4.3	8.1	5.0	8.1	3.8	7.0	4.0	2.3	3.0
An	0.7	0.8	0.4	1.2	0.8	0.5	0.9	0.6	0.7	0.8	0.4	0.3	0.2	0.4	0.5	0.2	0.7	0.4	0.4	0.4	0.3	0.7	1.2	0.3
Ns	1.4	1.1	1.6	0.9	1.6	0.7	0.0	0.4	0.5	0.8	1.1	0.6	1.6	1.0	1.0	0.1	0.7	0.4	0.7	0.6	0.8	0.8	0.5	0.6
Qz	1.0	2.7	4.5	3.1	2.5	3.8	2.5	0.5	0.7	3.1	2.8	3.0	2.6	2.1	1.1	0.9	1.3	1.3	2.2	1.8	2.8	3.3	2.7	3.0

End-members determined by the method of Peterson (1989a). ec = experimental charge; mph = homogeneous, unzoned microphenocryst; ph-c = phenocryst core from zoned crystal; ph-r = phenocryst rim from zoned crystal.

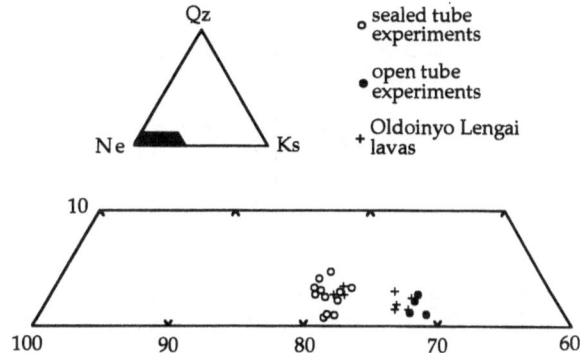

Fig. 2. Plot of nepheline analyses from experiments and homogenous, unzoned nephelines from wollastonite- and combeite-bearing nephelinites from Oldoinyo Lengai onto part of the nepheline-kalsilite-quartz ternary. Kalsilite-rich cluster of OL lava nephelines are from combeite nephelinite HOL6; remainder of OL lava nephelines from HOL14. All data from this chapter, and plotted in mol%

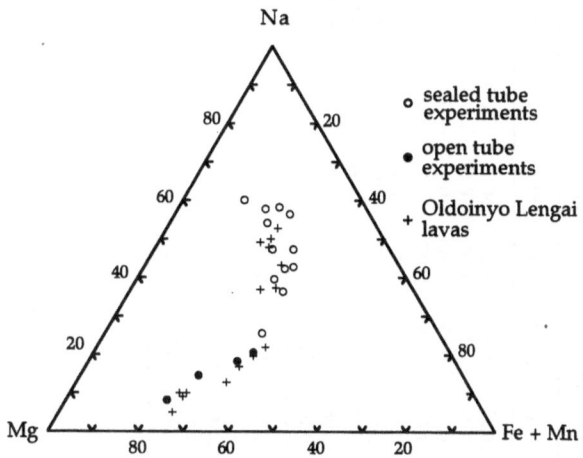

Fig. 3. Plot of clinopyroxene analyses from experiments and natural lavas and plutonic blocks from Oldoinyo Lengai onto the diopside (Mg)-hedenbergite (Fe^{2+} + Mn)-Aegirine (Na) ternary. Note that the clinopyroxenes from open-tube experiments are diopside-rich as compared to those from sealed-tube experiments. Sources of data for homogenous, unzoned clinopyroxene phenocrysts or microphenocrysts from natural rocks: Donaldson and Dawson (1978); Donaldson et al. (1987); Dawson et al. (1989); Peterson (1989a). Data are plotted in atomic%

ranging in composition from $Di_{39.6}He_{35.0}Ae_{25.4}$ to $Di_{19.3}He_{22.4}Ae_{58.3}$. At 750°C, clinopyroxenes from 100 MPa experiments are richer in diopside component than those from 300 or 400 Mpa experiments. However, at fixed pressure (100 MPa), clinopyroxenes from 750°C experiments are richer in diopside component than those from 700°C. Clinopyroxenes from sealed-tube experiments

Table 4. Representative microprobe analyses of clinopyroxene from experimental charges

Sample	BK411	BK412	BK413	BK425	BK428	BK431	BK424	BK427	BK430	BK436	BK429	BK437	BK426	BK432	BK417	BK423
SiO_2	51.01	50.38	51.00	50.15	50.53	50.97	50.35	50.36	51.56	50.99	50.52	51.04	49.58	51.00	50.64	51.47
TiO_2	0.80	0.70	0.90	0.51	0.40	0.75	0.65	0.66	0.46	0.71	0.62	0.61	0.52	0.58	0.49	0.49
Al_2O_3	0.60	0.36	1.13	0.61	0.47	0.43	0.90	0.45	0.42	1.00	0.40	0.52	0.84	0.98	0.80	0.73
FeO_t	20.63	22.01	21.99	21.34	21.91	22.96	17.28	20.37	21.83	21.63	22.24	23.29	16.25	10.8	16.72	12.61
MnO	0.58	0.72	0.70	0.41	0.74	0.73	0.45	0.64	0.63	0.49	0.48	0.66	0.45	0.32	0.41	0.42
MgO	5.08	4.23	3.65	4.02	3.61	3.14	6.35	4.90	4.11	4.34	3.88	2.90	7.79	11.47	7.51	10.18
CaO	15.58	14.27	13.26	12.91	12.71	11.98	19.31	15.45	13.68	15.28	12.52	12.10	20.22	23.04	18.92	20.96
Na_2O	4.76	5.88	6.18	6.97	6.96	7.30	3.12	5.00	5.73	5.35	6.79	7.10	2.26	1.04	2.69	1.93
K_2O	0.07	0.08	0.08	0.10	0.10	0.11	0.09	0.06	0.10	0.05	0.07	0.15	0.01	0.04	0.06	0.01
Total	99.04	98.55	98.81	96.92	97.33	98.26	98.41	97.83	98.42	99.79	97.45	98.22	97.91	99.23	98.18	98.79
Mg	29.7	26.3	21.5	26.1	22.9	19.3	39.6	29.9	23.7	26.2	23.9	17.7	48.6	69.4	44.1	59.2
Fe+Mn	34.1	26.3	31.1	14.9	19.5	22.4	35.0	30.5	33.4	31.7	21.8	25.8	33.1	22.4	35.3	26.2
Na	36.2	47.4	47.4	59.0	57.6	58.3	25.4	39.6	42.9	42.1	54.3	56.5	18.3	8.2	20.6	14.6

Table 5. Representative microprobe analyses of melanite garnet from experimental charges and lava HOL14

Sample	14a.1	14a.2	14a.3	14a.4	14a.5	14a.6	14a.7	14b	14c	14d	BK411	BK412	BK413	BK425	BK428	BK431	BK424	BK427	BK430	BK436	BK429	BK437	BK417	BK426
Type	zph	zph	zph	zph	zph	zph	zph	mph	mph	mph	ec	ec	ec	ec	ec	ec	ec	ec	ec	ec	ec	ec	ec	ec
SiO_2	29.02	26.81	27.99	27.82	30.89	29.97	30.37	29.32	29.54	29.47	28.67	28.84	30.88	29.34	29.40	30.74	28.27	30.52	30.04	28.73	28.79	30.10	30.11	29.86
TiO_2	14.57	15.54	14.48	16.40	10.86	11.05	11.71	12.46	12.06	11.80	13.40	13.66	12.66	11.13	11.77	9.87	13.48	12.09	11.07	12.68	11.77	11.47	11.56	11.72
Al_2O_3	0.73	0.73	0.64	0.59	1.07	1.00	0.93	1.01	1.02	0.86	0.75	1.09	1.10	0.77	0.81	0.83	1.05	0.92	0.90	0.87	0.76	0.37	0.76	0.73
FeO_t	19.99	19.43	20.17	19.51	21.27	21.39	21.55	20.93	21.47	21.13	20.65	20.46	20.88	20.33	21.25	21.46	19.51	22.79	21.29	20.94	21.26	21.16	20.53	20.45
MnO	0.31	0.37	0.35	0.33	0.31	0.34	0.30	0.30	0.36	0.28	0.37	0.21	0.39	0.34	0.41	0.45	0.32	0.37	0.34	0.41	0.37	0.38	0.31	0.37
MgO	0.75	0.89	0.78	0.83	0.50	0.50	0.38	0.61	0.52	0.50	0.75	0.68	0.55	0.38	0.53	0.14	0.48	0.42	0.42	0.44	0.64	0.31	0.45	0.44
CaO	31.72	30.90	31.48	31.05	31.62	31.71	31.29	31.20	31.61	31.25	29.29	30.72	30.05	31.04	30.30	31.77	31.12	31.11	32.22	30.83	30.58	32.02	30.74	32.64
Na_2O	0.43	0.4	0.41	0.52	0.21	0.41	0.57	0.28	0.42	0.28	0.34	0.36	0.43	0.83	0.41	0.60	0.64	0.38	0.62	0.32	0.42	1.05	0.45	0.81
K_2O	0.02	0.03	0.00	0.01	0.00	0.01	0.10	0.02	0.00	0.00	0.10	0.13	0.12	0.22	0.13	0.15	0.14	0.06	0.16	0.14	0.10	0.19	0.04	0.12
Total	97.54	95.10	96.30	97.06	96.73	96.38	97.20	96.13	97.00	95.57	94.32	96.15	97.06	94.38	95.01	96.01	95.01	98.66	97.06	95.36	94.69	97.05	94.95	97.14

Note: ec = experimental charge; mph = homogenous, unzoned microphenocrysts (14b, 14c, 14d); zph = zoned phenocryst, spots represent traverse from core (14a.1) to rim (14a.7).

have compositions which overlap only the more sodic compositions reported from unzoned phenocrysts or microphenocryts from Oldoinyo Lengai lavas (see Fig. 3). In contrast, clinopyroxene from HOL14- and HOL6-series open tube experiments at 50 MPa are much richer in diopside component ($Di_{69.4}He_{22.4}Ae_{8.2}$ to $Di_{44.1}He_{35.3}Ae_{20.6}$), and overlap the range of sodium-poor clinopyroxenes reported from Oldoinyo Lengai.

Titanium-rich (melanite) garnet formed dodecahedral crystals. Analyses from experiments are listed in Table 5, along with new garnet analyses (microphenocrysts and a strongly zoned phenocryst) from lava HOL14. Garnet compositions from open- and sealed-tube experiments are similar, ranging from 11.07 to 13.66 wt% TiO_2 and lie in the middle of the known compositional range of Ti-bearing garnets from Oldoinyo Lengai (7.17–18.25 wt% TiO_2: Donaldson et al. 1987). This is illustrated in Fig. 4, a Ti-Al-Fe^{3+} (atomic) ternary diagram. All Fe has been arbitrarily recast as Fe^{3+} since recalculation and assignment of cations by charge balance is problematic, as both Fe and Ti can occur in two valence states in melanite (Huggins et al. 1976).

Vishnevite formed hexagonal prisms, and was observed in all sealed tube runs as well as one open-tube run (BK432) Representative partial microprobe analyses are listed in Table 6, along with analyses of three natural vishnevites from Oldoinyo Lengai. EDS spectra indicate S ≫ Cl.

Wollastonite formed slender acicular crystals in runs at low pressure (100–50 MPa) and temperature (750–700 °C). It was not observed in runs at P > 100 MPa. Microprobe analyses of experimental wollastonites are listed in Table 7, along with compositions of phenocrysts and microphenocrysts from Oldoinyo Lengai silicate rocks. Natural and experimental wollastonites have similar compositions, both exhibiting minor substitution of FeO and MnO for CaO.

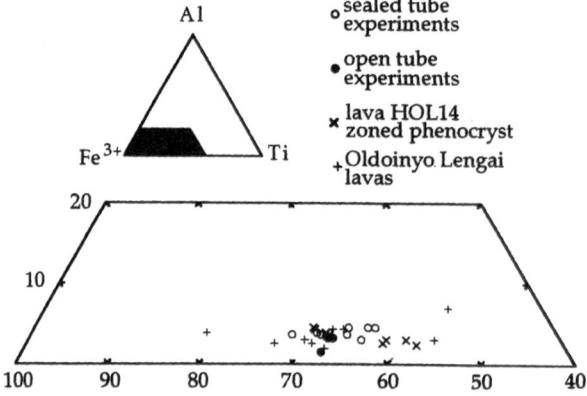

Fig. 4. Plot of melanite garnet analyses from experiments and natural lavas from Oldoinyo Lengai onto part of the Al-Fe^{3+}-Ti ternary. Sources of data for melanites (and schorlomites) from Oldoinyo Lengai (additional to this study): Huggins et al. (1976); Donaldson and Dawson (1978); Donaldson et al. (1988); Peterson (1989a). Data are plotted in atomic%

Table 6. Representative microprobe analyses of vishnevite from experimental charges and Oldoinyo Lengai samples

Sample Type	BK411 ec	BK425 ec	BK424 ec	BK430 ec	BK429 ec	D&D#1 hopper	D&D#2 verm	DDKBW ph
SiO$_2$	37.93	37.54	39.65	38.82	37.16	35.56	34.86	35.59
Al$_2$O$_3$	28.55	28.58	30.61	30.39	27.76	29.94	27.94	29.86
Fe$_2$O$_3$	1.42	1.49	3.69	1.56	1.24	1.01	2.20	0.27
CaO	1.80	1.37	2.03	1.32	0.74	1.70	1.76	4.57
Na$_2$O	20.00	19.51	17.18	19.37	20.87	21.03	21.55	15.79
K$_2$O	1.60	1.33	1.54	1.92	1.58	1.24	1.16	2.59
SO$_3$	det.na	det.na	det.na	det.na	det.na	5.19	6.93	6.97
Cl	0.79	det.na	det.na	det.na	det.na	1.80	0.99	1.00
Total	92.09	89.82	94.70	93.38	89.35	97.47	97.39	96.64

Sources of data: D&D = Donaldson and Dawson 1978; DDKBW = Donaldson et al. 1987. Note: ec = experimental charge; ph = phenocryst; hopper = hopper crystal; verm = vermicular crystal; det.na = determined semi-quantitatively by EDS only/not analyzed by WDS.

Table 7. Representative microprobe analyses of wollastonite from experimental charges and lava HOL14

Sample Type	BK427 ec	BK436 ec	BK429 ec	BK437 ec	DSS#5 ph	DSS#7 ph	P89a ph	HOL14-1 ph	HOL14-2 mph
SiO$_2$	50.39	50.88	50.00	49.61	51.73	52.03	50.77	51.06	50.94
TiO$_2$	0.12	0.07	0.07	0.06	0.00	0.00	n.a.	0.01	0.05
Al$_2$O$_3$	0.09	0.03	0.14	0.21	0.53	0.00	0.03	0.03	0.04
FeO$_t$	0.86	0.95	1.85	1.10	1.81	0.96	0.94	0.88	1.20
MnO	0.31	0.38	0.48	0.33	1.20	0.34	0.00	0.04	0.42
MgO	0.09	0.00	0.00	0.00	0.00	0.21	0.14	0.06	0.08
CaO	46.74	45.87	45.21	47.18	48.03	47.80	46.61	47.21	46.52
Na$_2$O	0.08	0.05	0.30	0.30	0.00	0.10	0.05	0.08	0.07
K$_2$O	0.18	0.09	0.24	0.20	0.21	0.01	0.02	0.03	0.13
Total	98.86	101.49	101.49	101.49	103.51	101.49	101.49	99.40	99.45

Sources of data: DDS = Dawson et al. 1989; P89a = Peterson 1989a. Note: ec = experimental charge; ph = phenocryst; mph = microphenocryst.

Combeite, leucite and sanidine were observed only in open-tube experiments (with the exception of sanidine in sealed-tube run BK429) and representative analyses are listed in Table 8. Combeite formed crystals with rhombohedral cross sections, and exhibited minor substitution of FeO$_t$. Leucite, see as octagonal cross sections, was near stoichiometric KAlSi$_2$O$_6$ in composition with substitution of 2.47 wt% FeO$_t$. Sanidine formed slender, elongated crystals, which were quite iron rich (3.53–4.18 wt FeO$_t$), similar to those in the glassy groundmass of HOL14.

Table 8. Representative microprobe analyses of sanidine, combeite and leucite from experimental charges

Sample Phase	BK426 Sanidine	BK429 Sanidine	BK432 Combeite	BK417 Combeite	BK423 Leucite
SiO$_2$	62.94	61.14	50.27	50.54	54.46
Al$_2$O$_3$	16.46	18.60	0.02	0.05	21.52
FeO$_t$	4.93	2.82	1.64	2.13	2.47
CaO	0.26	0.77	24.39	25.46	0.00
Na$_2$O	3.53	2.18	22.79	21.47	0.25
K$_2$O	11.18	13.06	0.13	0.17	20.59
Total	99.30	98.57	99.24	99.82	99.29

3.2 Liquid Phases

3.2.1 Silicate and Carbonate Liquids from Sealed-Tube Experiments

Table 9 lists compositions of quenched silicate and carbonate liquids from sealed-tube HOL14-series experiments. A sharp meniscus separates quenched silicate and carbonate liquids (Figs. 1, 2, 3a, 3b). Immiscible carbonate liquids quenched to dendritic intergrowths (Figs. 1, 2), similar to those described in previous studies (e.g. Wyllie and Tuttle 1960; Koster van Groos and Wyllie 1966; Freestone and Hamilton 1980). Carbonate liquid analyses are not listed for every run because of the difficulty in obtaining reliable analyses on small (30–50 µm) immiscible carbonate globules in some charges. The silicate liquids are all peralkaline, with molar (Na+K)/Al ranging from 1.49 to 2.14 and compositions are similar to those from the previous studies of Hamilton and co-workers. An important observation of this study is that peralkaline nephelinites have low liquidus temperatures (≈825 °C at 380 MPa) and are only 50% solidified at 700 °C and 100 MPa. This, however, is not surprising since the alkali and volatile contents of these melts are quite high.

Partitioning of major elements ($K_{D_{s/c}}$) between silicate and carbonate liquids in this study are comparable to the results of Freestone and Hamilton (1980) and Hamilton et al. (1989). For all P-T-X conditions, SiO$_2$, TiO$_2$, Al$_2$O$_3$ and FeO$_t$ partitioned into the silicate liquid, while Na$_2$O, CaO, P$_2$O$_5$, SrO, BaO, F and Cl partitioned into the carbonate liquid. MnO and MgO partitioning is variable, but near unity; these elements are at low concentration levels in both silicate and carbonate liquids, (generally <0.50 wt%). K$_2$O partitioning is variable, with $K_{D_{s/c}}$K$_2$O ranging from 0.83–2.70; the lower values are from 100 MPa experiments. Results for K$_2$O partitioning from Freestone and Hamilton (1980) and Hamilton et al. (1989) at comparable pressure ranged from 0.65–1.24. It is difficult to make direct comparisons between these data sets due to large temperature (1150–1000 °C vs. 750–700 °C) differences. Results from all studies, however, illustrate that K$_2$O partitioning into the carbonate liquid is enhanced at lower pressures and higher temperatures.

Table 9. Compositions of quenched silicate and carbonate liquids from sealed tube series experimental charges

Sample Phase	BK398 LS	BK399 LS	BK400 LS	BK400 LC	BK411 LS	BK412 LS	BK412 LC	BK413 LS	BK413 LC	BK425 LS	BK428 LS	BK428 LC	BK431 LS	BK431 LC	BK424 LS	BK427 LS	BK430 LS	BK430 LC	BK436 LS	BK429 LS	BK437 LS	BK437 LC
SiO_2	43.15	40.35	40.11	2.10	46.98	48.45	5.15	47.66	1.26	48.17	50.08	1.77	47.58	1.78	49.41	45.58	46.31	1.35	48.13	48.63	48.11	1.89
TiO_2	1.03	0.92	0.88	0.06	0.40	0.41	0.36	0.46	0.09	0.35	0.42	0.20	0.57	0.13	0.41	0.39	0.56	0.06	0.49	0.52	0.46	0.10
Al_2O_3	15.53	14.45	14.49	0.51	16.32	16.74	1.58	15.74	0.28	17.18	17.67	0.59	16.54	0.18	15.47	13.24	13.13	0.09	14.17	13.85	13.22	0.26
FeO	8.35	8.05	7.48	1.41	5.16	5.38	2.45	5.63	1.03	4.63	4.53	1.79	4.88	0.94	6.61	7.22	7.22	1.09	5.84	6.35	7.51	0.54
MnO	0.31	0.32	0.26	0.15	0.26	0.25	0.28	0.27	0.35	0.22	0.20	0.52	0.20	0.34	0.30	0.32	0.30	0.16	0.27	0.28	0.27	0.24
MgO	0.85	0.80	0.75	0.48	0.21	0.24	0.19	0.25	0.18	0.18	0.16	0.31	0.21	0.25	0.27	0.38	0.43	0.19	0.17	0.12	0.14	0.11
CaO	5.92	6.28	5.51	26.81	1.53	1.48	20.10	1.60	24.28	1.03	0.99	22.64	1.00	23.76	1.35	2.82	2.38	16.37	1.76	1.89	2.25	15.68
Na_2O	10.61	10.99	12.02	16.86	10.88	11.30	17.22	12.41	20.08	12.24	11.09	18.45	11.40	22.42	10.40	12.73	12.30	26.33	11.63	12.00	12.10	30.08
K_2O	5.19	5.22	5.39	2.03	7.23	7.21	3.27	7.38	2.73	7.07	7.41	3.43	7.56	3.51	7.24	6.84	7.33	4.92	7.21	7.27	7.16	8.65
P_2O_5	0.25	0.32	0.23	2.06	0.08	0.09	0.97	0.08	1.55	0.05	0.08	1.00	0.07	1.05	0.08	0.10	0.16	1.22	0.08	0.08	0.08	1.02
F	0.26	0.38	0.33	1.34	0.57	0.44	0.83	0.34	2.55	0.42	0.27	1.91	0.25	1.61	0.49	0.42	0.49	1.44	0.77	0.31	0.26	1.88
Cl	0.32	0.41	0.37	1.31	0.45	0.31	0.33	0.34	1.71	0.37	0.33	1.40	0.27	1.17	0.33	0.52	0.44	1.55	0.39	0.38	0.30	1.93
BaO	0.21	0.29	0.25	1.08	0.41	0.22	1.60	0.32	1.98	0.29	0.17	2.51	0.17	1.23	0.32	0.31	0.24	1.45	0.59	0.48	0.34	1.51
SrO	0.33	0.38	0.34	1.54	0.27	0.18	2.56	0.08	2.31	0.24	0.12	2.55	0.07	1.87	0.27	0.24	0.16	1.61	0.30	0.18	0.13	2.19
Total	92.31	89.16	88.41	57.74	90.75	92.70	56.89	92.56	60.38	92.44	93.52	59.07	90.77	60.24	92.95	91.11	91.45	57.83	91.80	92.34	92.33	66.08
$(Na+K)/Al$	1.49	1.64	1.78	58.67	1.58	1.58	20.16	1.80	128.50	1.62	1.49	57.71	1.63	225.90	1.61	2.14	2.14	540.20	1.90	1.99	2.00	226.20

Table 10. Compositions of quenched silicate liquids from open tube series experimental charges, plus green glasses from lavas HOL6 and HOL14

Sample	HOL-6	BK417	BK423	HOL-14	BK426	BK432
Type	Lava	ec	ec	Lava	ec	ec
Combeite	Yes	Yes	Yes	Yes	No	Yes
SiO_2	45.35	47.44	46.06	50.52	51.89	45.07
TiO_2	2.03	2.29	2.01	1.40	1.56	2.06
Al_2O_3	4.86	5.66	4.79	7.30	7.30	4.08
FeO	15.88	14.83	15.34	11.94	11.94	13.96
MnO	0.64	0.76	0.70	0.48	0.66	0.55
MgO	0.64	0.39	0.57	0.49	0.40	1.00
CaO	2.36	2.71	2.42	2.68	2.75	3.34
Na_2O	14.34	12.11	13.66	11.94	12.76	13.45
K_2O	8.34	6.86	8.62	6.50	6.20	7.41
P_2O_5	0.51	0.27	0.41	0.27	0.29	0.68
F	1.03	0.67	1.02	0.69	0.71	1.13
Cl	0.61	0.45	0.69	0.41	0.23	0.88
BaO	0.89	0.93	0.85	0.65	0.56	0.63
SrO	0.51	0.61	0.51	0.46	0.60	0.60
Total	97.99	95.98	97.65	97.44	97.85	94.84
(Na+K)/Al	6.71	4.83	6.64	4.13	3.79	7.39

Note: ec = experimental charge. Combeite (yes/no) indicates the presence or absence of combeite in that sample.

3.2.2 Silicate Liquids from Open-Tube Experiments

Compositions of quenched silicate liquids from open-tube HOL14- and HOL6-series experiments are listed in Table 10. Also listed in Table 10 are new microprobe analyses of the green silicate glasses from lavas HOL14 and HOL6. As previously noted by Peterson (1989a) these glasses are exceptionally peralkaline (4.13 and 6.71, respectively) and rich in halogens. Silicate liquids from open-tube experiments have compositions quite similar to the green glasses of the lavas; most notable are the high alkali, FeO_t and halogen levels and very low Al_2O_3 contents.

4 Discussion

4.1 The Immiscible Versus Conjugate Liquid Problem

A major drawback to models involving liquid immiscibility is whether the melts, if immiscible, are also conjugate. Twyman and Gittins (1987) have suggested that the occurrence of immiscibility in the Freestone and Hamilton (1980) experiments proved nothing and was an artifact of melting rocks with sharply divergent liquidi (natrocarbonatite $\approx 500\,°C$; silicate $\approx 1000\,°C$), i.e. the liquids

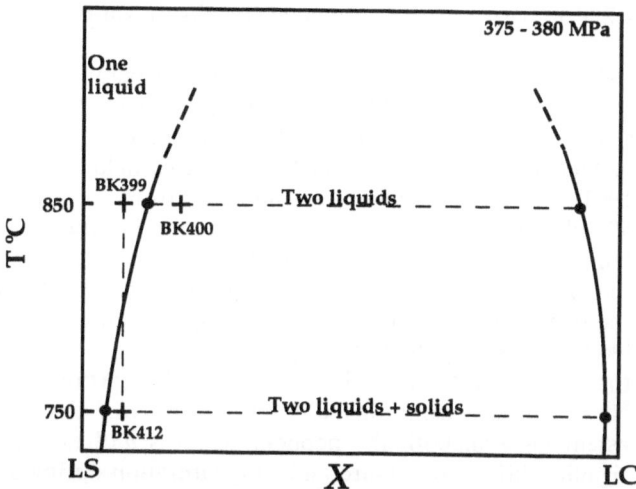

Fig. 5. Schematic diagram illustrating the relationship between bulk composition, temperature and the one liquid/two liquid field boundary. *LS* Silicate liquid; *LC* carbonate liquid; bulk composition denoted by a *cross*; conjugate liquids denoted by *filled circles* and joined by *dashed lines*. See text for details

are immiscible but do not represent a conjugate pair because they did not previously exist as a homogenous, high-temperature liquid. Both LeBas (1981, 1987, 1989) and Twyman and Gittins (1987) noted that the carbonate liquids of the Freestone and Hamilton experiments were exsolved at temperatures 350–550 °C hotter ("superheated") than the liquidus temperature of natrocarbonatite. However, this apparent "superheat" is easily attributed to the high carbonate mass fractions used in the early experiments.

In this study, bulk compositions utilized have much lower carbonate mass fractions. The relationship between bulk composition, temperature and the one-liquid/two-liquid field boundary is illustrated by the 380 MPa runs BK399, BK412 (HOL14*-5) and BK400 (HOL14*-10) as a schematic diagram in Fig. 5. BK400, with high carbonate mass fraction, lies inside the two liquid field; for this run both the silicate and carbonate liquid are "superheated" (not saturated with at least one solid phase). In contrast, at the same P-T conditions, BK399 (HOL14*-5), consists of one homogenous liquid. Run BK412 (two liquids + solids) illustrates that the two liquid experiments of this study show not only immiscibility, but also that the two melts represent conjugate pairs. Note that the only difference between BK399 and BK412 is lower T; P-X is constant and for this bulk composition at higher temperature (BK399) there is one homogenous liquid which exsolves a second liquid on cooling (BK412). By definition the immiscible carbonate liquid in BK412 is not "superheated" as it is saturated with melanite garnet and clinopyroxene.

4.2 Comparison of Experimental Results to Natural Silicate Lavas

Calculated eruption temperatures for highly peralkaline silicate lavas at Oldoinyo Lengai range from 500 to 775 °C, using the nepheline thermometers of Hamilton (1961) and Peterson (1989a). The occurrence of subhedral pyrite phenocrysts in HOL14 suggests eruption temperatures below 743 °C (Toulmin and Barton 1964). In general, phase assemblages from experiments at 700–750 °C and 50–100 MPa are consistent with those found in the lavas and support calculated temperature estimates. The major difference between the phase assemblages of the experiments and the lavas is the presence of a vishnevite group mineral in the experiments whereas sodalite + pyrite is present in the lavas. This is most likely due to slightly higher f_{O_2} in the experiments, i.e. $S + (2)O_2 \Rightarrow SO_4$; note that both sodalite and vishnevite have 1:1 Si:Al ratios and high Na (20–27 wt% Na_2O). Apart from this difference, one can distinguish two distinct phase assemblages in both the nephelinite lavas *and* experiments. Each assemblage signifies different pressure and CO_2 saturation during differentiation.

4.3 Differentiation of Peralkaline Nephelinites Saturated in CO_2 at Low P_{Total}

In experiments at 100 MPa, 700–750 °C and saturated in CO_2 (i.e. high p_{CO_2}), alkali-rich carbonatitic liquids are conjugate to silicate liquids of high, but near-constant peralkalinity (1.99–2.14: Table 9). The two liquids are in equilibrium with nepheline, clinopyroxene, melanite and vishnevite ± wollastonite, with the composition of these phases similar to those in the lavas (except vishnevite, see previous paragraph). Similarly, experiments using low peralkalinity lavas from Shombole (Kjarsgaard 1990; Kjarsgaard and Peterson 1991) produced two liquid plus solid assemblages, but carbonate melts were low alkali in composition and conjugate silicate liquids of low peralkalinity (1.20–1.30). The higher peralkalinity values reported for the Oldoinyo Lengai silicate liquids are related to the bulk compositions utilized. In both studies, silicate liquid peralkalinity remains nearly constant at fixed pressure and decreasing temperature. Results from the Shombole study illustrated that silicate liquid peralkalinity is buffered during isobaric cooling due to coupled solid phase precipitation and carbonate liquid exsolution. Any tendency to increased silicate liquid peralkalinity as a result of solid phase fractionation is negated by the changing composition of the exsolved carbonate liquid with decreasing temperature ($K_{D_{s/c}}Al_2O_3$ increases, $K_{D_{s/c}}Na_2O$ decreases and $K_{D_{s/c}}K_2O$ remains near-constant; Kjarsgaard 1990).

For Oldoinyo Lengai peralkaline nephelinite magmas exsolving alkali-rich carbonate liquids, residual silicate liquids should be buffered with respect to peralkalinity (as seen in experiments). Therefore extrusives forming from such magmas should have glasses with $(Na+K)/Al \approx 2.15$ (but with lower bulk rock peralkalinity $\approx 1.4–1.6$; due to the presence of solid phases of lower peralkalinity). However, the presence of small (10–30 μm) combeite and sanidine crystals in hyperalkaline glass with $(Na+K)/Al=3.95$ in wollastonite nephelinite lava HOL14 suggests limited very low pressure differentiation after carbonate liquid exsolution.

An exceptionally important additional point in terms of two-liquid relationships is the K_2O-rich (6.35–8.40 wt%) nature of natrocarbonatite. Since partitioning of K_2O into the carbonate liquid is favoured by lower pressures (<200 MPa) and higher temperatures (<1000 °C), the calculated and measured low eruption temperatures of both strongly peralkaline nephelinites and natrocarbonatite (<775 °C) demands low pressure exsolution ($\geqslant$35 MPa $\leqslant$100).

4.4 Differentiation of Peralkaline Nephelinites Undersaturated in CO_2 at Very Low P_{Total}

Combeite-phyric nephelinites, with bulk rock peralkalinities of 1.80–2.38, all contain hyperalkaline silicate glasses with (Na+K)/Al ranging from 4.13–6.71 (Peterson 1989a; Table 10, this chapter). One atmosphere experiments on the stability and precipitation of combeite (Peterson 1987, 1989b) indicate that combeite only forms when silicate liquid peralkalinity reaches minimum levels of 4.28 to 5.38. Results from open-tube experiments which were undersaturated with respect to CO_2 (this chapter) are consistent with this idea; combeite did not appear as a liquidus phase until a minimum peralkalinity of 3.95 was reached (e.g. Fig. 4). Glass analyses from experiments and natural lavas all support the notion that combeite is only stable in silicate liquids with abnormally high peralkalinity (>4).

While it can be argued that the 1 atm and open-tube experiments are not realistic, since p_{CO_2} was essentially zero (whilst the natural magmas were in equilibrium with a mixed H_2O/CO_2 vapour; as evidenced by highly vesicular lavas), nevertheless, the results are informative. Both sets of experiments illustrate residual liquid trends for silicate melts which are not in equilibrium with an immiscible carbonate liquid. This point is important, as silicate liquid peralkalinity is buffered below approximately 2.15 when an alkali-rich carbonate melt is exsolved, and at these peralkalinity levels combeite will not precipitate. Thus, it is suggested that combeite and natrocarbonatite are an incompatible phase assemblage, although one could argue that these two phases are compatible at P-T conditions not studied here. In this respect, note that Dawson et al. (1989) attributed the formation of combeite to the mixing of nephelinite and natrocarbonatite magma in the ash cloud during eruption.

Further understanding of this problem lies in comprehending the complex relationship between P_{Total}, CO_2 saturation and the one liquid/two-liquid field boundary. Koster van Groos and Wyllie (1966) provide an informative discourse on the subject. They illustrated that a stable two-liquid field depends on whether the vapour-saturated liquidus surface intersects the two-liquid region (see Fig. 6). Koster van Groos and Wyllie (1966) suggested that variation in CO_2 saturation level of a liquid is controlled by a series of pressure dependent decarbonation reactions. It is also known that CO_2 solubility in a silicate liquid is pressure dependent (e.g. Stolper et al. 1987; Pan et al. 1991), with solubility decreasing at lower pressures. The net observed effect is that with lower pressure, the silicate/carbonate two-liquid field decreases in width, as seen in the experiments of Koster van Groos and Wyllie (1966) and Freestone and Hamilton

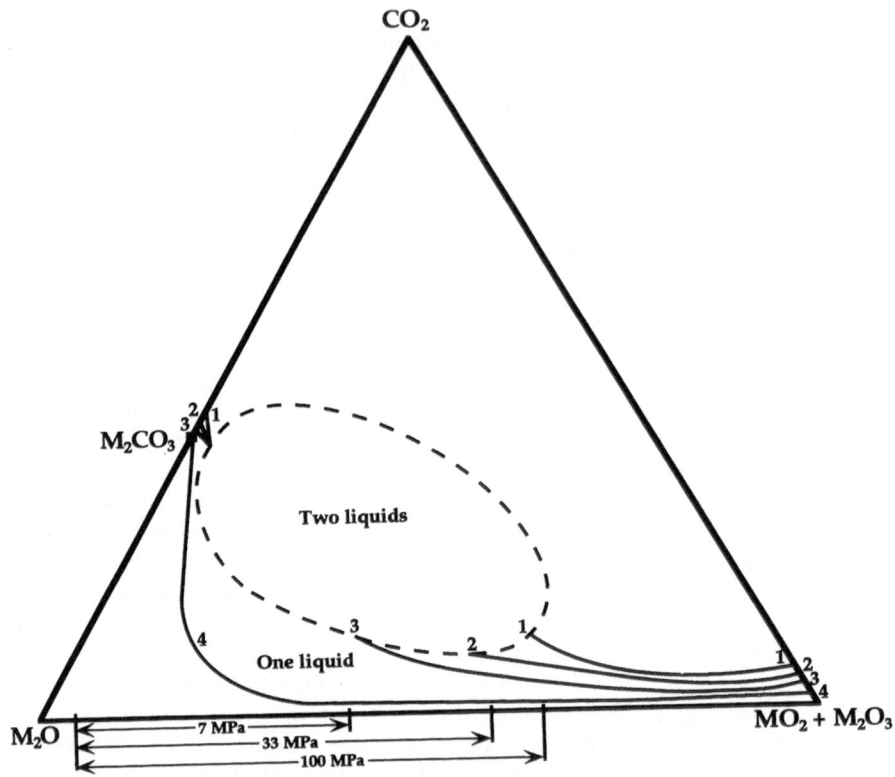

Fig. 6. Schematic diagram illustrating the intersection of the CO_2-saturated liquidus (*heavy solid lines* labelled *1* to *4*) and the two-liquid field (*dashed line*). With increasing pressure, CO_2 solubility increases and the vapour-saturated liquidus moves "up" in CO_2 space, as shown by the *heavy lines* labelled *4* (lowest pressure) to *1* (highest pressure). Note that line *4* does not intersect the two liquid field (1 atm) and that lines *3*, *2* and *1* (7, 33 and 100 MPa, respectively) intersect larger volumes of the two liquid field. This increased width of the two-liquid field is shown at the base of the triangle ($M_2O - MO_2 + M_2O_3$), a projection from CO_2 through the intersection of the vapour-saturated liquidus and the two-liquid field. Note that the vapour saturated liquidus for *1* to *3* (omitted for clarity) follows the dashed line delineating the two liquid field. (After Koster van Groos and Wyllie 1966, Fig. 10a–c). $M_2O = Na_2O + K_2O$; $MO_2 = SiO_2 + TiO_2$; $M_2O_3 = Al_2O_3 + Fe_2O_3$

(1980). We suspect this is mainly due to the lower CO_2 content of the silicate liquid.

The study of Koster van Groos and Wyllie (1966) illustrated a radical decrease in the size of the two-liquid field at very low pressures (between 33 and 7 MPa), and these results have important petrological implications for the differentiation of carbonated silicate liquids. A schematic P-X prism (Fig. 7), based on results from Koster van Groos and Wyllie (1966), Freestone and Hamilton (1980) and this study illustrates the salient aspects. Any silicate liquid, coexisting with an immiscible carbonate liquid (on the silicate limb of the solvus) with decreased pressure will no longer remain on the solvus, but lie in the one liquid (silicate) field. Further differentiation is then controlled by precipitation

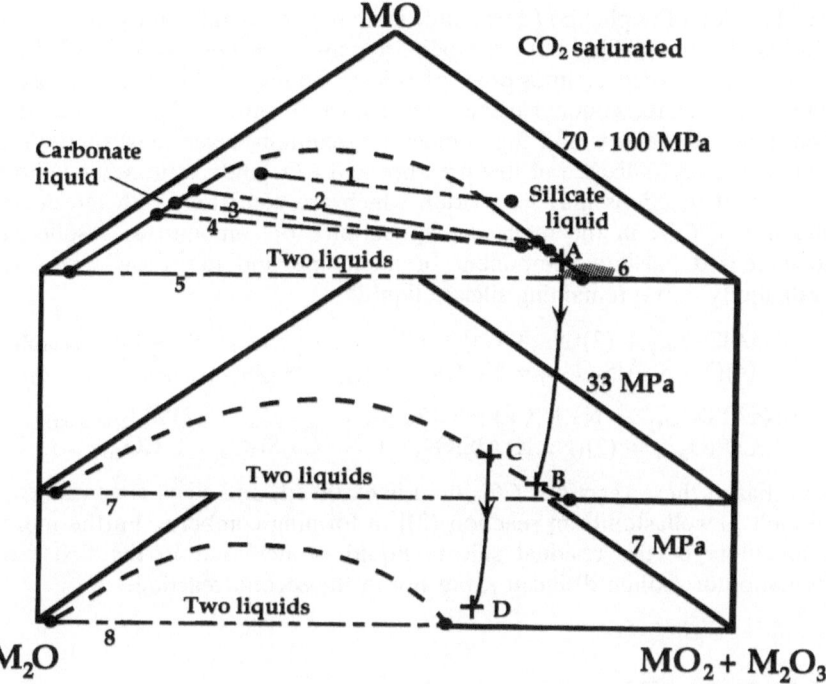

Fig. 7. Polythermal pressure-composition prism illustrating the decreasing size of the two liquid field with decreasing pressure for very alkali-rich systems. Conjugate liquid pairs are joined by *long/short dashed lines*. Two liquid pairs are: *1* and *4* = 70 MPa and 1100 °C, from Freestone and Hamilton, 1980; *2* and *3* = 100 MPa, 750 and 700 °C, this study; *5, 7* and *8* = 100, 33 and 7 MPa, respectively, and 900 °C, from Koster Van Groos and Wyllie (1966). *Shaded area* labelled *6* represents the compositions of all the silicate liquids at 100 MPa (which were in equilibrium with carbonate liquid) from the study of Koster Van Groos and Wyllie (1973). Note that for a silicate liquid at 100 MPa exsolving carbonate liquid (point *A*), with pressure decrease to 33 MPa (point *B*) and minor cooling, the residual liquid should still lie on the silicate limb of the solvus because the size of the two liquid field is little changed. However, a silicate liquid at point *C* (33 MPa), with pressure decrease to 7 MPa (point *D*), should cease to exsolve carbonate liquid due to a dramatic decreasing in the size of the two-liquid field. Continued differentiation of residual liquids which lie in the one-liquid (silicate) field (e.g. point *D*) at these very low pressures produces hyperalkaline residual liquids. $M_2O = Na_2O + K_2O$; $MO_2 = SiO_2 + TiO_2$; $M_2O_3 = Al_2O_3 + Fe_2O_3$; $MO = CaO + FeO + MgO + MnO$

of solid phases until the solvus is intersected again. While it is possible that the two liquid field could again be intersected with subsequent cooling, this scenario appears unlikely at very low pressure ($P_{Total} < 35$ MPa; see Fig. 7).

At low pressure ($P_{Total} < 35$ MPa), in the absence of an immiscible carbonate liquid, residual liquid peralkalinity will be controlled by the precipitation of solid phases. Comparing Figs. 3 and 4 (sealed- versus open-tube tube experimental charges), it can be seen that the relative proportions of precipitating solid phases is quite different, with enhanced nepheline precipitation in the open tube (CO_2 undersaturated charge). Since these liquids are peralkaline to start with, mass crystallization of nepheline ($\pm$sodalite) increases the peralkalinity of the residual liquid (Peterson 1989b), until combeite appears as a liquidus phase. The good congruence between compositions of silicate liquids (Table 10) and also solid phases (nepheline, clinopyroxene; see Tables 3 and 4; Figs. 2 and 3) from open-tube experiments and the combeite nephelinite lavas suggests the natural assemblages crystallizated at low pressure and CO_2 undersaturated conditions.

Potential reactions may be written which are consistent with the decreased solubility of CO_2 in the melt, the appearance of combeite as a solid phase, the increased kalsilite component of nepheline and lower SiO_2 plus higher peralkalinity in the remaining silicate liquid:

$$(4)NaAlSiO_{4(s)} + (2)(Na,K,Ca) - CO_{3(sil\ melt\ complexes)} + (4)SiO_{2(sil\ melt)}$$
$$\Rightarrow (4)(Na,K)AlSiO4_{(s)} + Na_2Ca_2Si_3O_{9(s)} + Na_2SiO_{3(sil\ melt)} + CO_{2(f)}. \quad (1)$$

$$(2)NaAlSiO_{4(s)} + (Na,K,Ca) - CO_{3(sil\ melt\ complexes)} + (2)SiO_{2(sil\ melt)}$$
$$+ CaSiO_{3(s)} \Rightarrow (2)(Na,K)AlSiO4_{(s)} + Na_2Ca_2Si_3O_{9(s)} + CO_{2(f)}. \quad (2)$$

Note that in these reactions CO_2 fluid is exsolved, and silica is consumed from the melt [+wollastonite in reaction (2)] in forming combeite. Furthermore, the peralkalinity of the residual silicate liquid is increased in the first reaction (formation of sodium disilicate), but not in the second reaction.

4.5 Formation of Natrocarbonatite Magma

Immiscible alkali-rich carbonate liquid exsolved at 100 MPa, 750–700 °C are alkali-, Sr-, Ba-, P- and halogen-rich, have low but appreciable concentrations of Si, Ti, Al, Fe and Mg (Table 9) and are multiply saturated with ferromagnesian solids. Erupted natrocarbonatite ($\approx$600 °C) is $\approx$ 100 °C cooler than the experimentally inferred exsolution temperature, and contains only rare ferromagnesian phenocryst phases. This could be used to suggest that minor magmatic differentiation occurs in the formation of a natrocarbonatite from the parental (immiscible) carbonate liquid.

Initially, exsolved carbonate liquids are in equilibrium with ferromagnesian solid phases (aegirine-augite, melanite, magnetite, Fig. 3). One can envisage efficient fractionation of these ferromagnesian solids concurrant with the separation of exsolved carbonate liquids from their silicate liquid host. Formation of a discrete body of carbonatite magma, along with further fractionation of ferromagnesian solids is enhanced by the low viscosity of the carbonate liquid. Residual carbonate liquids produced by this process should be essentially devoid of Si, Ti, Al, Fe and Mg. Reported analyses of natrocarbonatite with negligible contents of these elements (Keller and Krafft 1990) are consistent with this idea. Combining 100 MPa carbonate liquid exsolution/separation temperatures of 750–700 °C with the known liquidus temperature of natrocarbonatite (655 °C at 100 MPa: Cooper et al. 1975) suggests that initial fractionation of ferromagnesian solids occurs over a 50–100 °C cooling interval.

Subsequent precipitation and fractionation of nyerereite over a 35 °C cooling interval (gregoryite joins nyerereite at 620 °C: Cooper et al. 1975) and then

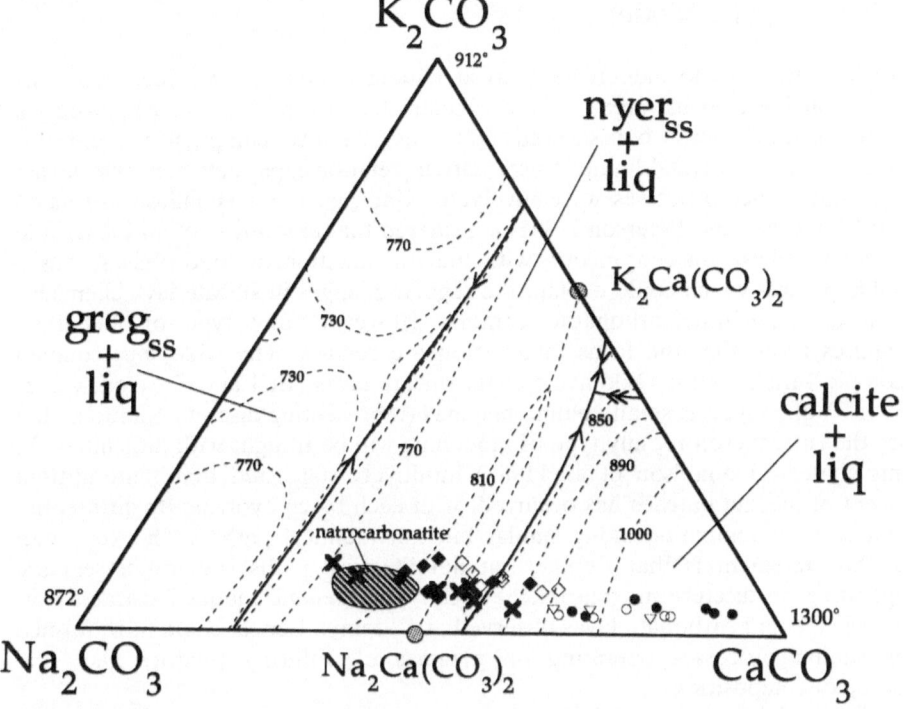

Fig. 8. Plot of immiscible carbonate liquids from this study (at 100–380 MPa) and previous experimental studies on Oldoinyo Lengai (at 300 MPa) and Shombole (at 500 MPa) onto the ternary $CaCO_3$-Na_2CO_3 K_2CO_3 100 MPa liquidus diagram of Cooper et al. (1975). Symbols: *crosses* Oldoinyo Lengai samples, this chapter; *filled diamonds*, Oldoinyo Lengai phonolite system (Hamilton et al. 1989); *open diamonds* Oldoinyo Lengai nephelinite system (Hamilton et al. 1989); *filled circles* Shombole SH49 + 10 wt% $CaCO_3$; *open circles* Shombole SH49; *open squares* Shombole SH40 (Kjarsgaard 1990). The *shaded area* labelled *natrocarbonatite* is from Peterson (1990). Data are plotted in wt%

gregoryite + nyerereite results in the formation of porphyritic natrocarbonatite. This fractionation sequence is supported by the observation that natrocarbonatitie lava compositions are distributed along a nyerereite subtraction line (see Fig. 8), corresponding to the predicted down temperature path on the liquidus surface (Peterson 1990). Precipitation of gregoryite + nyerereite produces a residuum (groundmass) exceptionally enriched in alkalies, Sr, Ba, and halogens. This is consistent with the composition of the lowest temperature (493 °C) natrocarbonatite liquids observed as being a filter-pressed aphyric residua from higher temperature porphyritic flows (Keller and Krafft 1990).

5 Petrogenetic Model

Previous attempts to model the lavas at Oldoinyo Lengai have been problematic. Donaldson et al. (1987) noted regular changes in the bulk compositions of the silicate lavas, but showed that mass balance computations did not support simple crystal-liquid fractionation relationships between the lavas, suggesting other processes were involved. Kjarsgaard and Hamilton (1989a,b) and Kjarsgaard and Peterson (1991) considered that exsolution of an immiscible carbonate phase, in conjunction with the fractionation of solid phases, was a viable process which could explain the regular changes in silicate lava chemistry seen at nephelinite/carbonatite centers. However, any type of modelling requires proof that the lavas in question are related. The wide, but coupled isotopic variations for Oldoinyo Lengai silicate lavas (Bell and Dawson, Chap. 7, this Vol.) suggests small volume magmas (representing discrete mantle melts) are the rule; therefore, any type of modelling will be fraught with difficulties. In this context, Donaldson et al. (1987) intuitively suggested that "intermittent ascent of magma batches has occurred, with each batch evolving by differentiation and contamination along similar but not identical paths". The key point of this statement is that similar, but not identical paths (i.e. processes) are operating on discrete magma batches. Our petrogenetic model illustrates how the silicate and carbonate lavas observed at Oldoinyo Lengai could have formed by similar processes operating on magmas of differing (major, trace and isotopic) composition.

The model is illustrated in Fig. 9. The magmatic lineage at Oldoinyo Lengai inferred to be associated with the carbonatites are olivine melilitites of low to moderate peralkalinity (the "Lengai Trend" of Peterson 1989a,b; see also Peterson and Kjarsgaard, Chap. 11, this Vol.). This magma forms by low degrees of partial melting of a volatile-rich lithosphere, and is carbonated (CO_2-bearing). Cooling and fractionation of solid phases increases the peralkalinity (and carbonate content) of the residual liquids, producing "fractionated nephelinites" typical of nephelinite/carbonatite complexes. Further cooling and fractionation increases the CO_2 content of the silicate liquid until saturation in this component occurs, the two liquid field is intersected and a carbonate liquid is exsolved. Most alkali-rich carbonatites at Oldoinyo Lengai have compositions quite similar to immiscible liquids produced in experiments, suggesting they have undergone little differentiation after exsolution.

The first type of carbonatite at Oldoinyo Lengai to consider are the nyerereite-phyric flows. Examples include GA47 (Deans and Roberts 1984; Dawson et al. 1987) and OL4 (Keller and Krafft 1990; Koberski and Keller, Chap. 6, this Vol.). These trachytic textured flows are dominated by nyerereite phenocrysts and microphenocrysts (which are in the process of altering to gaylussite and then pirssonite) in a matrix of apatite, opaques, fluorite and secondary calcite. Gregoryite is considered to be absent (Deans and Roberts 1984; Keller and Krafft 1990; Koberski and Keller, Chap. 6, this Vol.), and the inferred occurrence of rare, primary calcite cores in nyerereite (Dawson et al. 1987) has not been confirmed. The absence of both calcite and gregoryite in these samples lead both Deans and Roberts (1984) and Keller and Krafft (1990) to suggest that

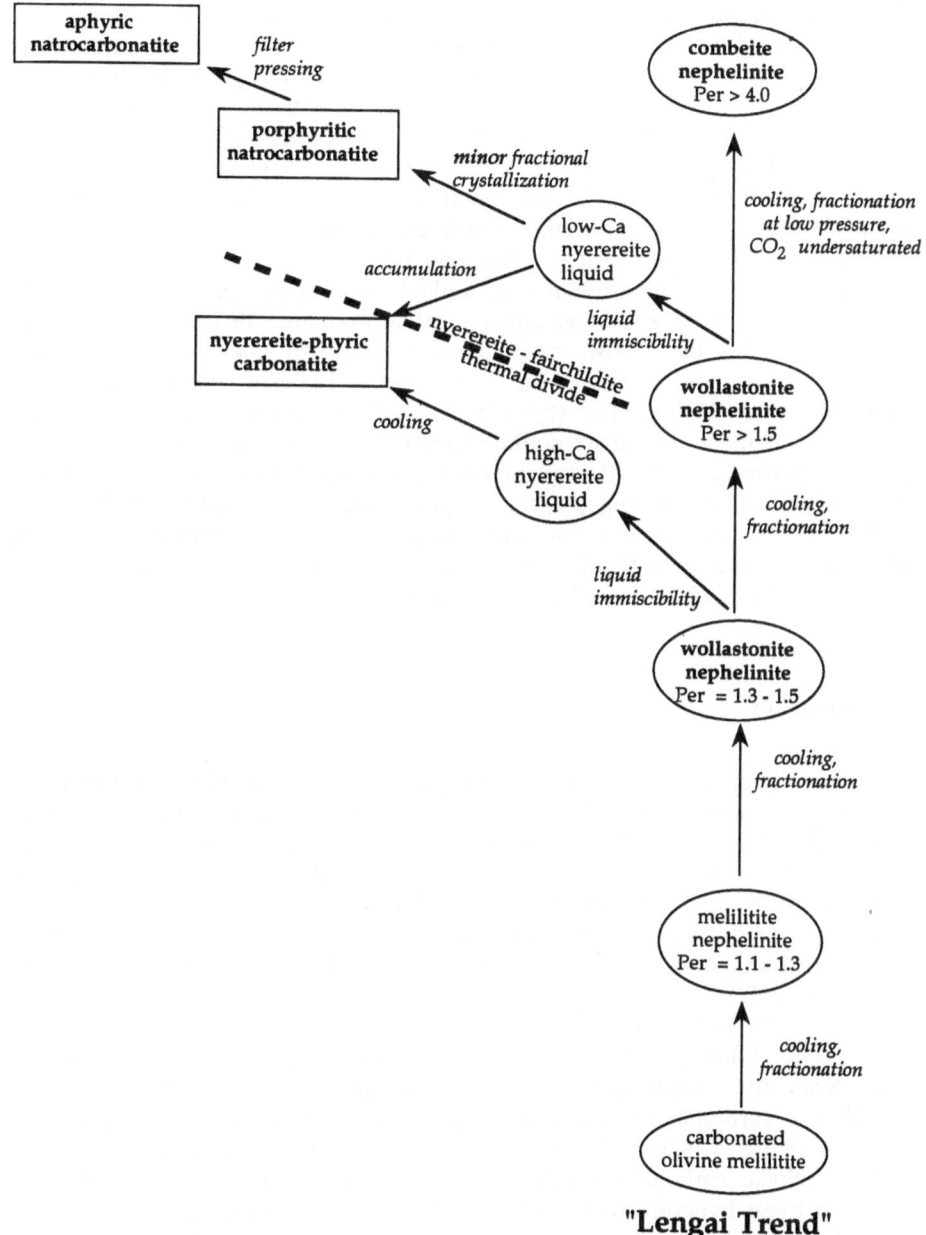

Fig. 9. Flow sheet illustrating the probable relationships between associated high alkali carbonate and hyperalkaline silicate lavas at Oldoinyo Lengai. Silicate lavas with low peralkalinity ("Shombole Trend"; typical of other nephelinite-carbonatite centres in East Africa), which are also found at Oldoinyo Lengai, are considered to be of another magmatic lineage and not included in the flow sheet. Any occurrence (if found) of low alkali carbonatite at Oldoinyo Lengai is inferred to be associated with "Shombole Trend" lavas. Per = peralkalinity

the parent liquid composition lay between the calcite-nyerereite cotectic and the nyerereite-fairchildite thermal divide (see Fig. 8). However, an additional possibility exists, in which the nyerereite-phyric flows are cumulates from carbonate liquids whose composition lies on the Na-rich side of the nyerereite-fairchildite thermal divide (see Fig. 8). For either parent carbonate liquid, the conjugate silicate liquid which exsolved it must have been at least moderately peralkaline ("Lengai trend"); low peralkalinity ("Shombole Trend") silicate liquids exsolve carbonate liquids which all lie in the calcite stability field (Kjarsgaard 1990; Kjarsgaard and Peterson 1991).

The second (and most intensively studied) type of carbonatite at Oldoinyo Lengai are the porphyritic and aphyric natrocarbonatite flows. The carbonate parent liquid for these flows must lie on the Na-rich side of the nyerereite-fairchildite thermal divide (see Figs. 8, 9). Jago and Gittins (1991) have demonstrated that very high fluorine (>8wt% F) calciocarbonatite can differentiate to natrocarbonatite via a calcite-liquid reaction which occurs in the presence of fluorite. However, natural natrocarbonatite has much lower F (<4wt%) and fluorite is a late crystallizing groundmass phase (Keller and Krafft 1990; Peterson 1990). We see no evidence for calcite-liquid reactions, or fluorite phases, in these or any other magmatic carbonatites, and therefore conclude that high alkali carbonate liquids form by liquid immiscibility.

6 Summary

Exceptionally alkalic carbonate parent liquids can be exsolved by liquid immiscibility from highly peralkaline carbonated wollastonite nephelinite magmas which are CO_2 saturated at relatively low P_{Total} (between 35 and 100 MPa). These alkalic carbonate parent liquids are exsolved at temperatures between 750 and 700 °C and are not superheated. Subsequent separation of exsolved carbonate liquids from their hosts, coupled with minor differentiation (fractionation of ferromagnesian solids plus precipitation of nyerereite and gregoryite) results in the formation of porphyritic natrocarbonatite. The rare, exceptionally nyerereite-rich carbonatites are suggested to be cumulates from this differentiation process. Filter pressing (as suggested by Keller and Krafft 1990) of residual melt from the porphyritic natrocarbonatite produces the aphyric variety. Peralkaline carbonated wollastonite nephelinite magma which differentiate at very low pressure (≤35 MPa) and are understurated with respect to CO_2 do not exsolve alkalic carbonate liquids. Resultant combeite precipitation can only occur in these hyperalkaline silicate liquids if the peralkalinity ≥4. Combeite nephelinites are expected to be found associated with natrocarbonatite, but the conjugate liquid to natrocarbonatite is suggested to be a wollastonite nephelinite.

Acknowledgements. This work was initated while BAK was a PDF at Manchester (NERC grant to DLH), and completed while a NSERC PDF at the Geological Survey of Canada. Reviews by P.J. Wyllie, A.F. Koster van Groos, K. Currie and K. Bell improved the manuscript. This is GSC contribution #36592.

References

Cooper AF, Gittins G, Tuttle OF (1975) The system Na_2CO_3-K_2CO_3-$CaCO_3$ at 1 kilobar and its significance in carbonatite petrogenesis. Am J Sci 275:534–560

Dawson JB (1962a) Sodium carbonate lavas from Oldoinyo Lengai, Tanganyika. Nature 195:1075–1076

Dawson, JB (1962b) The geology of Oldoinyo Lengai. Bull Volcanol 24:349–387

Dawson JB (1989) Sodium carbonate extrusions from Oldoinyo Lengai, Tanzania: implications for carbonatite complex genesis. In: Bell K (ed) Carbonatites – genesis and evolution. Unwin Hyman, London, pp 255–277

Dawson JB, Garson MS, Roberts B (1987) Altered former alkalic lavas from Oldoinyo Lengai, Tanzania: implications for calcite carbonatite lavas. Geology 15:765–768

Dawson JB, Smith JV, Steele IM (1989) Combeite $(Na_{2.33}Ca_{1.74}others_{0.12})Si_3O_9$ from Oldoinyo Lengai, Tanzania. J Geol 97:365–372

Deans T, Roberts R (1984) Carbonatite tuffs and lava clasts of the Tinderet foothills, western Kenya: a study in calcified natrocarbonatites. J Geol Soc Lond 141:563–580

Donaldson CH, Dawson JB (1978) Skeletal crystallization and residual glass compositions in a cellular alkalic pyroxenite nodule from Oldoinyo Lengai. Contrib Mineral Petrol 67:139–149

Donaldson CH, Dawson JB, Kanaris-Sotiriou R, Batchelor RA, Walsh NJ (1987) The silicate lavas of Oldoinyo Lengai, Tanzania. Neues Jahrb Min Abh 156:247–279

Droop GTR (1987) A general equation for estimating Fe^{3+} concentrations in ferromagnesian silicates and oxides from microprobe analyses, using stoichiometric criteria. Min Mag 51:431–435

Freestone IC, Hamilton DL (1980) The role of liquid immiscibility in the genesis of carbonatites: an experimental study. Contrib Mineral Petrol 73:105–117

Hamilton DL (1961) Nephelines as crystallization emperature indicators. J Geol 69:321–329

Hamilton DL, Freestone IC, Dawson JB, Donaldson CH (1979) Origin of carbonatites by liquid immiscibility. Nature 279:52–54

Hamilton DL, Bedson P, Esson J (1989) The behaviour of trace elements in the evolution of carbonatites. In: Bell K (ed) Carbonatites – genesis and evolution. Unwin Hyman, London, pp 405–427

Huggins FE, Virgo D, Huckenholz HG (1976) The crystal chemistry of melanites and schorlomites. Carnegie Inst Wash Year Book 75:705–711

Jago BC, Gittins J (1991) The role of fluorine in carbonatite magma evolution. Nature 349:56–58

Keller J, Krafft M (1990) Effusive natrocarbonatite activity of Oldoinyo Lengai, June 1988. Bull Volcanol 52:629–645

Kjarsgaard BA (1990) Nephelinite-carbonatite genesis: experiments on liquid immiscibility in alkali silicate-carbonate systems. PhD Thesis, University of Manchester

Kjarsgaard BA, Hamilton DL (1989a) Melting experiments on Shombole nephelinites: silicate/carbonate liquid immiscibility, phase relations and the liquid line of descent. Geol Assoc Can/Mineral Assoc Can Progr with Abstr 14:A50

Kjarsgaard BA, Hamilton DL (1989b) The genesis of carbonatites by immiscibility. In: Bell K (ed) Carbonatites – genesis and evolution. Unwin Hyman, London, pp 388–404

Kjarsgaard BA, Peterson TD (1991) Nephelinite-carbonatite liquid immiscibility at Shombole volcano, East Africa: petrographic and experimental evidence. Min Petrol 43:293–314

Koster van Groos AF, Wyllie PJ (1963) Experimental data bearing on the role of liquid immiscibility in the genesis of carbonatites. Nature 199:801–802

Koster van Groos AF, Wyllie PJ (1966) Liquid immiscibility in the system Na_2O-Al_2O_3-SiO_2-CO_2 at pressures up to 1 kilobar. Am J Sci 264:234–255

Koster van Groos AF, Wyllie PJ (1968) Liquid immiscibility in the join $NaAlSi_3O_8$-Na_2CO_3-H_2O. Am J Sci 266:932–967

Koster van Groos AF, Wyllie PJ (1973) Liquid immiscibility in the join $NaAlSi_3O_8$-$CaAl_2Si_2O_8$-Na_2CO_3-H_2O. Am J Sci 273:465–487

LeBas MJ (1981) Carbonatite magmas. Min Mag 44:133–40

LeBas MJ (1987) Nephelinites and carbonatites. In: Fitton JG, Upton BGJ (eds) Alkaline igneous rocks. Blackwell, London, pp 53–85

LeBas MJ (1989) Diversification of carbonatite. In: Bell K (ed) Carbonatities – genesis and evolution. Unwin Hyman, London, pp 428–447

Morimoto N (1988) Nomenclature of pyroxenes. Min Petrol 39:55–76

Pan V, Holloway JR, Hervig RL (1991) The pressure and temperature dependence of carbon dioxide solubility in tholeiitic basalt melts. Geochim Cosmochim Acta 55:1587–1595

Peterson TD (1987) The petrogenesis and evolution of nephelinite-carbonatite magmas. PhD Thesis, Johns Hopkins University, Baltimore (unpublished)

Peterson TD (1989a) Peralkaline nephelinites I. Comparative petrology of Shombole and Oldoinyo Lengai, East Africa. Contrib Mineral Petrol 101:458–478

Peterson TD (1989b) Peralkaline nephelinites II. Low pressure fractionation and the hypersodic lavas of Oldoinyo Lengai. Contrib Mineral Petrol 102:336–346

Peterson TD (1990) Petrology and genesis of natrocarbonatite. Contrib Mineral Petro 105: 143–155

Stolper E, Fine G, Johnson T, Newman S (1987) Solubility of carbon dioxide in albitic melt. Am Mineral 72:1071–1085

Toulmin P III, Barton PB Jr (1964) A thermodynamic study of pyrite and pyrrhotite. Geochim Cosmochim Acta 28:641–671

Twyman JD, Gittins J (1987) Alkalic carbonatite magmas: parental or derivative? In: Fitton JG, Upton BGJ (eds) Alkaline igneous rocks. Blackwell, London, pp 85–94

Wyllie PJ, Tuttle OF (1960) The system CaO-CO$_2$-H$_2$O and the origin of carbonatites. J Petrol 1:1–46

Experimental Constraints
on the Possible Mantle Origin of Natrocarbonatite

R.J. Sweeney[1], T.J. Falloon[2], and D.H. Green[3]

Abstract

The phase relations of a "primary" sodic dolomitic carbonatite (CM1), determined by Wallace and Green (1988) to be in equilibrium with an amphibole lherzolite assemblage, have been investigated from 5 to 27 kb over a temperature range of 650–1225 °C with 2 and 4 wt% added water in order to test the model, first proposed by Wallace and Green (1988), that natrocarbonatite may result from the closed system crystal fractionation of such a "primary" composition. Our experiments suggest that if Oldoinyo Lengai magmas evolve from a primary carbonatitic liquid at depth by crystallization of dolomite, olivine and spinel, then this liquid must be more silicic, less FeO-rich and have a greater K/Na ratio than the Wallace and Green (1988) "primary" carbonatite composition. These suggested differences may be consistent with differences in the mantle source region from which the Oldoinyo Lengai composition was derived and the peridotite composition (Hawaiian Pyrolite) with which the Wallace and Green (1988) composition was equilibrated.

The phase relationships of CM1 also demonstrate the existence of a decarbonation reaction to lower pressures and higher temperatures which results in the formation of a CO_2-rich fluid phase. This fluid phase contains alkali elements and chlorine and may be responsible for the fenitization of wall-rocks associated with many carbonatite complexes. Additionally, a late-stage fractionate would be extremely sodic (approximately 40 wt% Na_2O) and may also fenitize wall-rocks.

A twofold increase in the added water content (from 2 to 4 wt%) depressed the liquidus temperature by <30 °C and had little effect on the observed liquidus phase relationships. No evidence for liquid immiscibility was observed in the P-T region studied.

1 Introduction

Wallace and Green (1988) established that a sodic dolomitic carbonatite melt coexists with an amphibole lherzolite assemblage in a P-T field of 21–31 kb and 930–1080 °C (composition CM1 in Table 1). The melt composition is unlike natural extrusive calcitic carbonatites (Keller 1981, 1989) or natrocarbonatite (e.g. Keller and Krafft 1990). However, Wallace and Green (1988) calculated a

[1] Present address: Institut für Mineralogie und Petrographie der ETH, ETH-Zentrum, 8092 Zürich, Switzerland
[2] Department of Geology, University of Tasmania, GPO Box 252C, Hobart, Tasmania 7001, Australia
[3] Research School of Earth Sciences, Australian National University, Canberra 2601 ACT, Australia

Table 1. Compostions of "primary" carbonatite, natural natrocarbonatite, and selected experimental melts

	CMI[a]	Old. Len.[b]	CMI+4 wt% water									CMI+2 wt% water								
			T3372	T3404	T3427	T3400	T3403	T3406	T3373	T3428	T3405	T3351	T3349	T3397	T3192	T3438	T3393	T3439	T3444	T3408
Pressure (kb)			27	22	15	15	15	10	10	5	5	27	22	20	15	15	15	10	10	5
Temp (°C)			100	1000	1050	1000	850	900	800	850	750	1000	1000	900	1125	1000	850	1000	800	750
Vol.melt (%)			66	67	98	94	1	68	7	40–60	1	50	65	1	100	83	4	91	10	3
No. anal.			(n = 4)	(n = 3)	(n = 3)	(n = 4)	(n = 4)	(n = 4)	(n = 3)	(n = 3)	(n = 3)	(n = 4)	(n = 4)	(n = 3)	(n = 4)	(n = 1)	(n = 4)	(n = 3)	(n = 3)	(n = 4)
wt%																				
SiO_2	3.23	0.24	3.75	3.05	2.22	1.77	0.40	0.73	1.27	4.50	0.56	1.66	1.69	0.56	4.79	1.81	<0.05	1.58	9.86	<0.05
TiO_2	0.53	0.03	0.65	0.46	0.46	0.41	0.56	0.20	0.13	0.59	0.22	0.45	0.41	0.17	0.73	0.18	0.18	0.29	0.68	0.26
Al_2O_3	2.14	–	2.47	1.81	0.64	0.70	2.76	1.51	1.64	2.59	0.36	0.89	0.65	0.67	3.11	1.08	0.11	1.37	6.22	<0.05
Cr_2O_3	0.24		0.02	0.04	<0.05	<0.05	<0.05	<0.05	<0.05	0.31	0.09	0.03	<0.05	<0.05	0.35	<0.05	<0.05	0.14	0.68	<0.05
FeO	5.06	0.42	4.59	4.35	4.54	3.49	3.04	2.81	1.54	5.35	1.09	5.43	4.81	3.07	5.64	4.55	1.20	3.89	7.01	0.32
MgO	15.49	0.57	11.22	10.63	14.13	12.48	1.12	13.39	13.80	17.63	1.03	10.17	11.85	1.01	17.65	14.02	0.53	13.53	12.43	0.51
CaO	23.07	12.33	15.23	16.40	22.93	18.95	1.77	18.25	9.65	17.88	3.41	15.54	18.75	2.68	20.62	23.11	1.09	23.09	10.12	0.17
Na_2O	5.51	31.08	6.09	9.12	7.28	7.95	43.09	10.25	23.24	7.06	40.87	12.68	6.55	50.79	2.76	6.12	45.08	8.13	10.57	48.43
K_2O	0.39	9.61	0.45	0.47	0.34	0.43	0.61	0.43	0.69	0.50	0.14	0.33	0.32	0.32	0.35	1.63	0.38	1.20	0.85	0.21
P_2O_5	0.53	0.73	0.86	0.75	0.66	0.68	2.15	0.97	3.36	0.73	0.38	0.94	0.86	1.72	0.57	0.67	0.88	0.69	1.41	0.55
Cl	0.66	5.30	0.19	0.60	0.93	1.67	0.49	0.86	2.62	2.51	0.20	0.82	nd	0.10	0.29	1.54	0.24	2.02	1.18	0.17
SO_3	0.88	4.92	<0.05	0.15	<0.05	<0.05	0.29	0.11	0.05	0.21	0.06	<0.05	nd	0.13	0.13	0.10	0.07	<0.05	0.17	<0.05
Total	57.73	65.23	45.51	47.83	54.13	48.52	56.27	49.50	57.99	46.3	48.40	48.92	45.89	61.24	56.99	54.81	49.79	49.79	61.18	50.69
Mg#	84.5	70.8	81.3	81.3	84.7	86.4	39.6	89.5	94.1	85.4	62.9	76.9	81.5	37.0	84.8	84.6	43.9	86.1	76.0	73.4

a "Primary" mantle carbonatite composition after Wallace and Green (1988). Also Contains 0.33% F, 2.03% H_2O, 1.61% SrO and 38.32% CO_2.
b Oldoinyo lengai aphyric natrocarbonatite, sample 104 from Keller and Krafft (1990).

Table 2. Experimental runs

Experiment	Pressure (kb)	Temp. (°C)	Duration (h. min.)	Melt vol% (type)[a]
CM1+2 wt% H_2O				
T3351	27	1000	66.00	45%(qch)
T3349	22	1000	73.00	65%(qch)
T3397	20	900	24.00	1%(Na)
T3509	18	960	47.00	55%(qch)
T3177	15	1225	2.00	100%(qch)
T3192	15	1125	2.30	100%(qch)
T3438	15	1000	2.20	83%(qch)
T3393	15	850	43.23	4%(Na)
T3439	10	1000	2.20	91%(qch)
T3444	10	800	73.15	10%(qch)
T3443	5	950	1.05	98%(qch)
T3408	5	750	24.40	3%(Na)
T3388	5	650	42.45	0%(ss)
CM1+4 wt% H_2O				
T3372	27	1000	24.00	66%(qch)
T3434	22	1120	2.05	100%(qch)
T3404	22	1000	42.10	67%(qch)
T3474	18	960	44.53	51%(qch)
T3427	15	1050	3.00	98%(qch)
T3400	15	1000	24.00	94%(qch)
T3403	15	850	31.40	1%(Na)
T3441	10	1000	1.40	100%(qch)
T3406	10	900	50.55	68%(qch)
T3413[b]	10	870	11.00	55%(qch)
T3373	10	800	61.40	7%(Na)
T3421	10	700	25.05	0%(ss)
T3428	5	850	2.00	40–60%(qch)
T3405	5	750	26.05	1%(Na)
T3418	5	650	20.10	0%(ss)
CM1+10% H_2O				
T3440	22	1000	5.10	77%(qch)

[a] Melt types: qch is quenched dolomite melt, Na is a sodic melt, ss is subsolidus.
[b] Capsule leaked.

Run products were analyzed using a Cameca SX50 wavelength dispersive 3-spectrometer microprobe at the University of Tasmania. Precisions were typically 3% (relative), and lower limits of detection typically 0.05%. Most analyses were undertaken at 15 kV and 20 nA, with the exception of some sodic melt pools, where filament current was decreased to 10 nA to limit potential Na loss due to volatilization.

We have considered the 4 wt% H_2O series of experiments in detail and compare these results with a more limited body of data on the 2% H_2O experiments (Table 2). Some runs are clearly vapour-saturated (CO_2 + H_2O fluid) at lower pressures and, in some instances, this ruptured the capsule on pressure release. Where textures are preserved, these leaked runs (T3433 at 5 kb and 950 °C, and T3413) are included with the results as they assist definition of a field of vapour saturation. Although the effect of additional water will be complex, it may be assumed that the 2% H_2O experiments will not have a lower liquidus or solidus or phase appearance temperatures at a given pressure than the 4% H_2O experiments.

3 Results

3.1 Solidus and Liquidus Positions

The solidus and liquidus positions for the 2 and 4% H_2O systems are depicted in Fig. 1a,b along with estimates of the amount of melt present. The textures

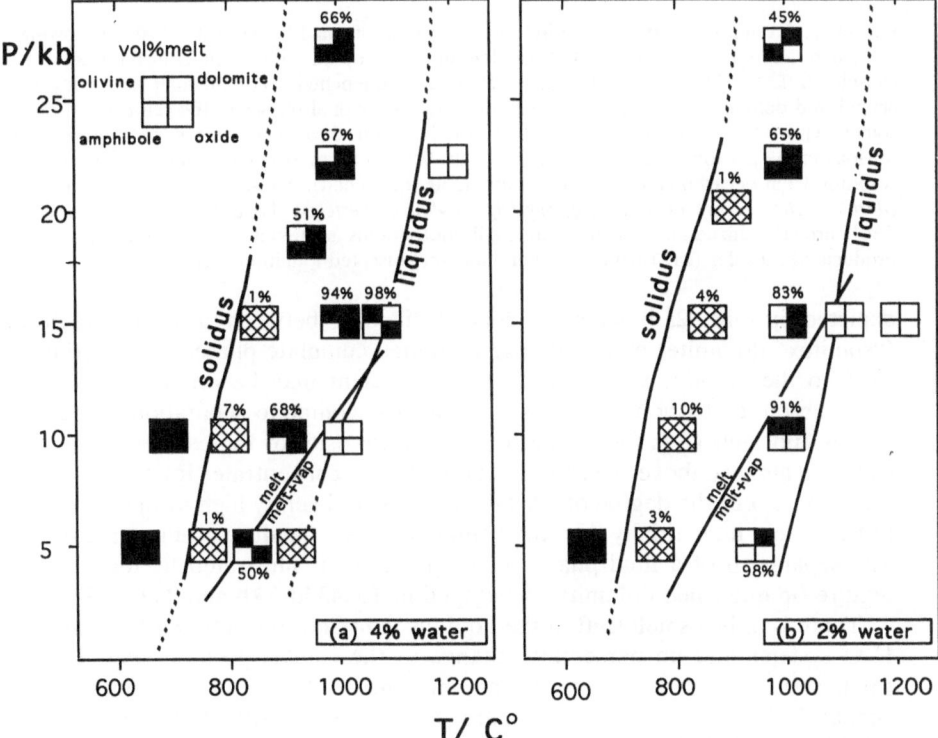

Fig. 1a,b. Solidus and liquidus positions for CM1 + 4 wt% H_2O (**a**) and CM1 + 2 wt% H_2O (**b**) and liquidus assemblages at various degrees of crystallization (percentage values refer to the volume of liquid present in each run). *Solid lines* are known boundaries and *dashed lines* are inferred boundaries. All runs on the higher-temperature side of the decarbonation line contain a vapour phase. *Cross-hatched runs* represent experiments that were too fine-grained to determine a residual mineralogy or in which a capsule breach occurred. Garnet is present in the 2% H_2O system only at 27 kb and oxide phases include spinel and ilmenite (Table 3)

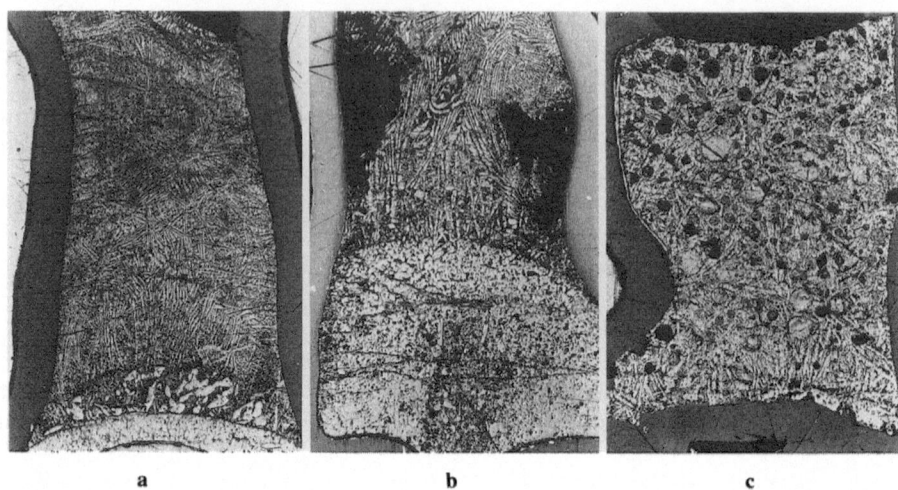

a   b   c

Fig. 2a–c. Examples of textures at different P-T conditions. **a** T3400 (15 kb, 1000 °C) showing a quenched dolomite melt with crystallized dolomite, olivine and minor spinel at the base of the capsule. **b** T3404 (22 kb, 1000 °C), again showing a quenched dolomite melt and a layer of crystallized dolomite and amphibole (e.g. the large hexagonal phases in the centre) and minor spinel. The different degree of crystallization between these two runs is pronounced and obvious in the photomicrographs. **c** T3443 (5 kb, 950 °C) as an example of the vesicular texture resulting from decarbonation of a dolomitic melt (quenched). These vesicles are now cavities (*dark vesicles*) or are filled with epoxy (*light-coloured vesicles*). Larger phases at the base of T3443 are crystallized magnesian calcites. All photographs are taken using reflected light of run products in capsules (ca. 1.6 mm internal diameter) bisected longitudinally

depicted in Fig. 2a,b show a clear distinction between a quench-textured ("spinifex" dolomite) melt and coarse-grained cumulate phases. In experiments close to the liquidus the amount of melt present may be estimated to within ±5% absolute (by volume) as there is almost complete separation of liquidus phases and melt (Fig. 2a,b) due to the low viscosity of the carbonate melt. In experiments just above the solidus, the melt also concentrates toward the top of the capsule, but the degree of crystallization is sufficiently high to cause this melt to be present interstitially, and the volume is thus difficult to estimate accurately. The separation of a fluid phase at low pressures is illustrated by the vesicular texture (in quenched dolomite) developed in T3443 at 5 kb and 950 °C (Fig. 2c).

There is only a small shift in the liquidus temperatures between the 2 and 4% H_2O systems and no perceptible change in the solidus position (Fig. 1). This small shift is emphasized by noting the similar degree of crystallization for similar P-T conditions (e.g. compare T3351 with T3372; T3349 with T3404; T3393 with T3403 in Table 2).

3.2 Mineralogy in the Melting Interval

Estimates of the modal abundances of the liquidus phases are only made for runs with appreciable melt (≥50 vol%, Table 3) as the fine-grained nature of the

Table 3. Visual volume estimates of melt and liquidus mineral abundances for runs with >50vol% melting

	(kb)	(°C)	Melt	dol[a]	mg-cc	calcite	oliv	amph	ga	sp	ilm
CM1+2 wt% H₂O											
T3349	22	1000	65	28				6		1	
T3351	27	1000	45	25				8	12		
T3438	15	1000	83	16							1
T3439	10	1000	91		8.5		0.5				
T3443	5	950	98			2					
T3509	18	960	55	40				5			
CM1+4 wt% H₂O											
T3372	27	1000	66	25				8		1	
T3400	15	1000	94	4		Trace	2			Trace	
T3404	22	1000	67	26				6			1
T3406	10	900	68	23			8			1	
T3427	15	1050		98				1.8			0.2
T3428	5	850	50		50						
T3474	18	960	51	42				7		Trace	Trace

[a] dol = dolomite, mg-cc = magnesium calcite (5–10% MgO); oliv = olivine; amph = amphibole; ga = garnet; sp = spinel; ilm = ilmenite.

nearer-solidus runs makes it difficult to identify residual phases and estimate modal proportions. Furthermore, the presence of probable residual calcite and magnesite crystals (from the starting mix) in near-solidus runs argues that it is unlikely that equilibrium was achieved over the duration of the experiment. In contrast, in experiments with $\geq 50\%$ melt (i.e. $T \geq 850\,°C$) the homogeneity of phases and the absence of residual starting mix components suggests equilibrium assemblages. Selected mineral compositions are given in Table 4.

For all conditions where an appreciable volume of melt is present, dolomite is the dominant phenocryst phase (Table 3). In the 2% H_2O system, garnet + amphibole were the major silicate phases at 27 kb, with amphibole + spinel at 22 kb. In the 4% H_2O system, however, the non-carbonate assemblage was dominated by amphibole + spinel at both 27 and 22 kb. At pressures less than 22 kb, the near-liquidus assemblage for the hydrous runs is dominated by dolomite with olivine and minor spinel (Table 3).

The $MgCO_3$ content of dolomite decreases with pressure (Table 4; Fig. 3). The Mg-number (Mg#, calculated assuming all Fe as Fe^{2+}) of liquidus dolomite and olivine correlates with the Mg# of the associated equilibrium melt, with both olivine and dolomite having Mg# greater than their eqilibrium melts (Tables 1 and 4). Fe/Mg mineral-melt exchange coefficients ($K_D = (Fe^{2+}/Mg^{2+})_{mineral} \cdot (Mg^{2+}/Fe^{2+})_{melt}$, with total Fe as Fe^{2+}) calculated for olivine were in the range 0.581–0.587 for the 4% H_2O system (three runs at pressures ≥ 10 kb) and 0.324–0.374 in the 2% H_2O system (T3439 at 5 kb). Fe-Mg partitioning between dolomite and melt was in the range $K_D = 0.38$–0.61.

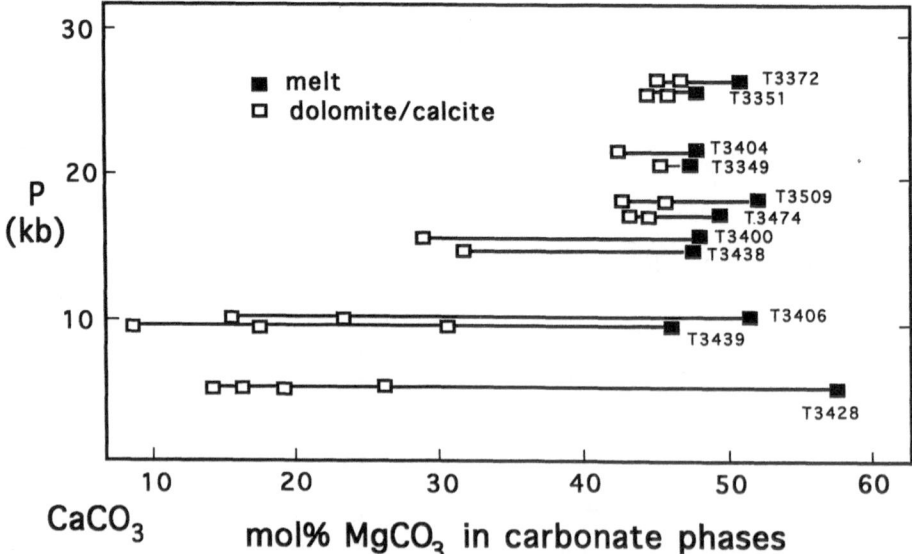

Fig. 3. Variation of molar % magnesite and calcite of crystallized carbonate minerals and associated melt for runs with more than 50% melt present. The change in the partitioning of the calcite-magnesite components with pressure is well demonstrated

Table 4. Selected mineral compositions[a]

Carbonates

	CM1+4 wt% water					CM1+2 wt% water			
Anal. no.	dol 1	dol 2		dol 2	dol 1	dol 1	dol 1	dol 1	Mg-cc2
Exp	T3372	T3404	T3427	T3400	T3406	T3351	T3349	T3438	T3439
P (kb)	27	22	15	15	10	27	22	15	10
T (°C)	1000	1000	1050	1000	900	1000	1000	1000	1000
wt%									
FeO	3.89	3.60		1.81	0.96	4.10	3.52	2.54	1.04
MgO	18.45	20.18		14.15	11.08	19.87	19.93	15.96	7.66
CaO	31.66	37.64		49.08	51.12	33.19	32.67	47.79	51.69
Na_2O	<0.05	0.11		0.06	0.08	0.09	0.13	0.09	0.22
P_2O_5	0.14	0.17		0.22	0.21	0.14	0.16	0.24	0.59
Total	54.14	61.70		65.32	63.45	57.39	56.41	66.62	61.20
Mg#	89.4	90.9		93.3	95.4	89.6	91.0	91.8	92.9

Silicates

	CM1+4 wt% water					CM1+2 wt% water			
Anal. no.	amph 1	amph 2	ol 1	ol 1	ol 6	ga 1	amph 1	amph 1	ol 2
Exp	T3372	T3404	T3427	T3400	T3406	T3351	T3351	T3349	T3439
wt%									
SiO_2	40.03	41.94	41.58	40.93	40.63	40.39	41.77	42.51	42.05
TiO_2	1.31	1.66	<0.05	<0.05	<0.05	0.72	1.60	1.90	<0.05
Al_2O_3	16.55	15.89	<0.05	<0.05	<0.05	21.73	16.54	15.11	<0.05
Cr_2O_3	1.00	0.67	<0.05	<0.05	<0.05	1.01	0.42	0.73	<0.05
FeO	6.39	6.97	9.45	8.18	6.34	11.72	7.39	4.87	5.68
MgO	14.27	15.01	50.21	50.35	51.45	16.25	14.07	16.87	52.83
CaO	7.76	7.95	0.09	0.16	0.35	7.86	7.47	9.58	0.34
Na_2O	4.80	4.84	<0.05	<0.05	<0.05	0.18	5.21	4.14	<0.05
K_2O	0.70	0.63	<0.05	<0.05	<0.05	<0.05	0.76	0.64	<0.05
Total	92.81	95.56	101.33	99.62	98.77	99.86	95.23	96.35	100.90
Mg#	79.9	79.3	90.4	91.6	93.5	71.2	77.2	86.1	94.3

Table 4. *Continued*

Oxides (Fe_2O_3 calc. on stoichiometry)	CM1+4 wt% water				CM1+2 wt% water	
	sp 1 T3372	sp 3 T3404	sp 1 T3427	sp 4 T3406	sp 1 T3349	ilm 1 T3438
SiO_2	<0.05	<0.05	0.39	0.08	<0.05	<0.05
TiO_2	0.21	0.26	1.20	0.66	0.21	49.65
Al_2O_3	40.89	45.35	41.04	53.56	52.24	1.21
Cr_2O_3	22.14	17.03	10.21	5.74	14.30	1.67
Fe_2O_3	4.18	5.11	12.27	8.68	1.05	7.76
FeO	14.58	12.14	11.34	8.59	12.98	20.36
MgO	15.17	17.24	16.93	20.80	17.65	13.63
CaO	0.08	0.13	2.55	<0.05	0.25	<0.05
Total	97.25	97.25	95.93	98.11	98.69	94.28
Mg#	65.0	71.7	72.7	81.2	70.8	54.4

[a] dol = dolomite; mg-cc = magnesium calcite; amph = amphibole; ol = olivine; ga = garnet, sp = spinel, ilm = ilmenite.

Amphibole is characteristically sodic and aluminous (Table 4) in both sets of runs and compositionally similar to magnesio-hastingsite. Hogarth (1989) notes that magnesio-hastingsite is the most common amphibole early in a carbonatite crystallization history.

3.3 Composition of the Melt

The dominant control on the composition of the melt is the degree of crystallization. For example, the compositions of melt at 27 and 22 kb (1000 °C) in the 4% H_2O series (Table 1) are quite similar and the degree of crystallization in both these runs is almost identical (32–35 vol%). As the solidus is approached melts become progressively more sodic, for example at 15 kb Na_2O increases from about 8% at 1000 °C to 45% at 850 °C, over a crystallization interval of approximately 90 vol%. Similarly, at 10 kb, Na_2O increases from 10% at 900 °C to 23% at 800 °C, over a crystallization interval of about 60 vol%.

It is significant that no evidence for liquid immiscibility in this P-T range was found. This is consistent with the conclusions of Baker and Wyllie (1990) in the range 20 to 30 kb and suggests that the one-liquid field of Kjarsgaard and Hamilton (1989, Fig. 15.1) should be expanded slightly to encompass the Wallace and Green (1988) "primary" carbonatite composition (see Figs. 1 and 2; Baker and Wyllie 1990).

The occurrence of small carbonate and chloride quench minerals in vesicles suggests that the fluid phase present in the near-liquidus experiments at 10 and 5 kb (e.g. Fig. 2c) and in above liquidus experiments at 15 kb, is CO_2- and chlorine-rich. A decarbonation reaction explains a dominance of CO_2 in the fluid and it is important to note that this fluid may also be enriched in Na and K. One run at (T3428, 5 kb, 850 °C) contained abundant NaCl crystals, commonly growing inward from the margins of vesicles, and in another instance (T3439, 10 kb, 1000 °C) a KCl quench phase occupied approximately 40% (by area) of a section through a vesicle. Exsolution of a fluid from a carbonatite melt at low pressure and high temperature may produce an alkali-enriched and chlorine-enriched $H_2O + CO_2$ fluid, and this may prove to be a mechanism for producing the fenitization of wall rocks associated with many carbonatites. Additionally, the late-stage highly sodic melt present in all near-solidus runs may also be a fenitizing agent.

4 Discussion

The increase in water content from 2 to 4 wt% in the starting bulk composition has little effect on the liquidus and solidus temperatures within the resolution of the experiments. The increase of H_2O to 4% can affect the liquidus mineralogy slightly. For example, amphibole is the only stable silicate at 27 kb and 1000 °C in the 4% H_2O system, wheares at this condition in the 2% H_2O system garnet occurs. The presence of garnet is consistent with the much lower Al_2O_3 content of the melt in the 2% H_2O system (Table 1). However, the residual mineralogy is very similar for the 2 and 4% H_2O systems at 22 kb (dolomite + amphibole + spinel) and consequently the melt compositions at this degree of melting (65–68%) are similar (Table 1). The crystallization of a generally more aluminous spinel in the 2% H_2O system may cause the ubiquitously lower melt Al contents compared to the 4% H_2O system (Tables 1 and 4).

A recent study by Jago and Gittins (1991) has shown that elevated F contents depress liquidus temperatures more effectively than increasing H_2O in a Na-Ca-K carbonate system. Although we have not measured F in our melt, it is present in the bulk composition (0.33% F) and, assuming perfect incompatibility, would comprise 3.3–6.6% in a 90–95% fractionate, which would have a liquidus temperature in the range 700–600 °C for an Oldoinyo Lengai composition at 1 kb (cf. Fig. 1, Jago and Gittins 1991). Thus, our results in the multi-component system are consistent with the experimental work of Jago and Gittins (1991) in the four-component system Na_2CO_3-Ca_2CO_3-K_2CO_3-F.

Our experiments suggest that the Wallace and Green (1988) composition is not a liquid at their specified conditions. It is saturated in dolomite, amphibole and spinel, which leaves a melt more enriched in Na and depleted in Mg and Ca relative to the starting composition (melts T3372 and T3351 in Table 1). The absence of saturation in peridotitic phases (olivine + orthopyroxene ± clino-pyroxene) under these conditions is problematic, although we suggest that relatively small modifications to bulk composition would generate multiple saturation in peridotitic phases. Considering the absence of any resolvable liquidus shift upon increasing water from 2 to 4 wt%, it is considered unlikely that any further increase in water content will depress the liquidus substantially. To substantiate this, one experiment was conducted with 8 wt% added water, for a total water content of 10 wt%. It showed 23 vol% crystallization, less than the 33 vol% crystallization observed for the 4 wt% H_2O system, but still substantially below the liquidus. This supports the results of Jago and Gittins (1991), which show that water is not the main agent in lowering the carbonatite liquidus.

The liquidus assemblage (4% H_2O system) of major dolomite > olivine > spinel persists over at least the crystallization interval 0–40 vol% (Table 3). This confirms the prediction of Wallace and Green (1988) that low-pressure frac-tionation would be dominated by this assemblage. The presence of dolomite as the dominant liquidus phase in runs where the proportion of melt is high (and the melt is calcitic) is consistent with the ubiquitous presence of this mineral in natural calcic carbonatites (e.g. Hogarth 1989; Cooper and Reid 1991; Reid and Cooper 1991). The ease with which dolomite/calcite crystallized in these experi-ments is explained by the presence of a dolomite-calcite eutectic near a dolomitic bulk composition in the CaO-MgO-CO_2-H_2O system (Fanelli et al. 1986; Wyllie 1989). Wyllie (1989) points out, too, that the co-precipitation of dolomite and calcite may occur over an extensive temperature interval (880–660 °C) at 2 kb (see Fig. 20.6 in Wyllie 1989).

An interesting and possibly important observation is the increasingly calcium-rich and magnesium-poor nature of the near-liquidus carbonates as pressure decreases (Fig. 3). This dependence on pressure is not associated with a change in melt composition as melts are uniformly dolomitic (45–58 mol% $MgCO_3$).

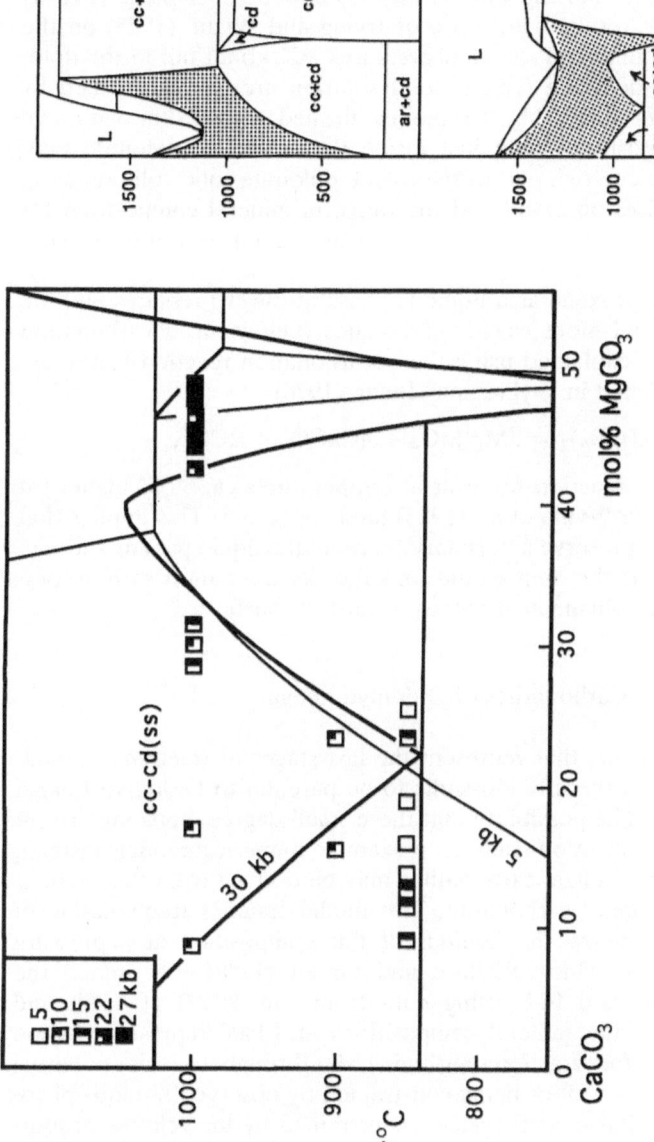

Fig. 4. Molar % magnesite and calcite in carbonate phases as a function of temperature and pressure indicating the different run pressures. The *shaded areas* in the *smaller figures* represent the regions of calcite-dolomite (cc-cd) solid solution in the end-member system CaCO₃-MgCO₃, as determined by Irving and Wyllie (1975, Fig. 6). Their phase boundaries at 30 and 5 kb are superimposed on the larger figure, which illustrates the results of this study. Abbreviations: *cm* magnesite; *ar* aragonite; *p* periclase; *L* liquid; *V* vapour

As the bulk system studied here is composed mostly of $CaCO_3$ and $MgCO_3$, we have projected the measured dolomite compositions from multicomponent space (e.g. addition of Na_2CO_3) onto the $CaCO_3$-$MgCO_3$ end-member plane (Fig. 4) in order to compare our results with those of Irving and Wyllie (1975) on the end-member system. Dolomites formed at pressures $\geqslant$22 kb all fall in the dolomite end of the calcite-dolomite (cc-cd) solid solution area as established by Irving and Wyllie (1975) at 30 kb. Dolomites formed at $\leqslant$15 kb, with two exceptions having compositions only just inside the calcite + dolomite two-phase field, fall in the calcite-rich part of the calcite-dolomite solid-solution area. Thus the carbonate phases observed and the range in mineral composition for any given pressure are consistent with available experimental data in the calcite-magnesite system.

The absence of clinopyroxene as a liquidus phase at lower pressures and the stabilization of olivine and more calcitic carbonates (below the decarbonation reaction, Fig. 1) may be explained using the decarbonation reaction below the calcite-dolomite solvus listed in Wyllie and Huang (1976)

$$CaMgSi_2O_6 + 3CaMg(CO_3)_2 = 2Mg_2SiO_4 + 4CaCO_3 + 2CO_2,$$

although we observe this reaction to occur at temperatures ca. 80 °C higher (at equivalent pressures) than Wyllie et al. (1983) later suggested. This implies that it would be impossible to preserve a "primary" carbonatite liquid passing through the lithosphere and crust; the melt would undergo decarbonation with accompanying crystallization of olivine as it moves toward the surface.

4.1 Comparison with the Carbonatites of Oldoinyo Lengai

The late-stage residual melts that represent the last stages of fractional crystallization in this study are generally too sodic to be parental to Oldoinyo Lengai compositions (Table 1). The possibility that these small-degree melts underwent considerable quench modification leads us to examine numerical models instead, to test whether Oldoinyo Lengai carbonatites may be derived from the Wallace and Green (1988) "primary" carbonatite. This model assumes fractionation of phases crystallized just below the liquidus of this composition at a pressure between 22 kb (the P at which Wallace and Green (1988) determined the "primary" composition) and 0 kb using data from run T3400 at 15 kb and 1000 °C. It is likely that the mineral compositions in T3400 represent a good aggregate composition for fractionating minerals throughout this potential crystallization interval. As well as being constrained by observed liquidus phase compositions, a mass-balance model is also constrained by the relative proportions in which they crystallize. The experiment T3400 contains coexisting dolomite, olivine and minor spinel with a ratio of dolomite/olivine of 2 (Table 2). The crystallization model to test for a relationship between Oldoinyo Lengai liquid and CM1 was calculated by fitting CaO and MgO (the major components of the melt) exactly (Table 5). Al_2O_3 is constrained by Cr-Al spinel crystallization. Thus, the Oldoinyo Lengai composition may be generated from the CM1 composition by 59 wt% crystallization of an assemblage comprising 18 wt%

Table 5. Crystallization model for an Oldoinyo Lengai magma: the exercise calculates a putative parent for Oldoinyo Lengai by adding olivine, dolomite and spinel in proportions required to match the CaO and MgO contents of CM1

Weight fraction %	ol1 0.184	dol2 0.367	Minerals sp4 0.041	+ Derivative melt[a] 0.408	= parent calc.	CM1
SiO_2	40.93	0.00	0.08	0.24	7.62	3.23
TiO_2	0.00	0.00	0.66	0.03	0.04	0.53
Al_2O_3	0.00	0.00	53.56	0.00	2.19	2.14
Cr_2O_3	0.00	0.00	5.74	0.00	0.23	0.24
FeO	8.18	1.81	16.40	0.42	3.01	5.06
MnO	0.00	0.00	0.00	0.00	0.00	0.00
MgO	50.35	14.15	20.80	0.57	15.53	15.49
CaO	0.16	49.08	0.00	12.33	23.09	23.07
Na_2O	0.00	0.06	0.00	31.08	12.71	5.51
K_2O	0.00	0.00	0.00	9.61	3.92	0.39
P_2O_5	0.00	0.00	0.00	0.73	0.30	0.53
NiO	0.00	0.00	0.00	0.00	0.00	0.00
LOI	0.00	0.00	0.00	0.00	0.00	0.00
Total	99.62	65.10	97.24	55.01	68.63	56.19
Mg#	91.6	93.3	69.3	70.8	90.2	84.5

[a] Oldinyo Lengai natrocarbonatite (Table 1).
dol = dolomite, sp = spinel, ol = olivine.

olivine + 37 wt% dolomite + 4 wt% spinel from the CM1 composition with some significant exceptions. To achieve a perfect numerical fit, SiO_2, Na_2O, K_2O, FeO and P_2O_5 would have to be adjusted in the "primary" composition. SiO_2, Na_2O and K_2O are too low in the "primary" composition by factors of 2.4, 2.3 and 10, respectively, and FeO and P_2O_5 would have to be reduced by a factor of 0.58 to obtain an exact numerical fit. TiO_2 may be explained by small amounts of ilmenite crystallization (ca. 1%). Most obvious is the increase in K/Na ratio of the "primary" composition by a factor of 4, which most likely reflects differences between the Oldoinyo Lengai peridotitic mantle source and that used by Wallace and Green (1988) (Hawaiian Pyrolite) to equilibrate with a carbonatite melt. Generation of an Oldoinyo Lengai primary composition (Mg# 90.2) from a more magnesian (FeO-depleted) mantle than the Wallace and Green (1988) composition (Mg# 84.5) is also implied by the numerical model (Table 5).

Extrapolation parallel to the liquidus and solidus (Fig. 1) for the CM1 composition for 59 wt% crystallization implies a magma temperature of about 750 °C at 0 kb. This is considerably above the value of 655 °C determined at 1 kb for the natrocarbonatite by Cooper et al. (1975) and the maximum of 544 °C measured in the lava lakes at Oldoinyo Lengai (Keller and Krafft 1990). The increase in K/Na suggested by the numerical model for the primary composition would most likely decrease the liquidus and solidus temperatures significantly, potentially to values consistent with eruption temperatures.

Only closed-system processes have been considered in the above numerical model and it is anticipated that the interaction of a dolomitic melt with a peridotic wall-rock will be significant. For instance, the reaction of a dolomitic melt with enstatite (an abundant mineral in peridotitic mantle) may produce a magnesite component (in the melt) and diopside:

$$Mg_2Si_2O_6 + CaMg(CO_3)_2 = CaMgSi_2O_6 + 2MgCO_3.$$
(enstatite)   (melt)   (diopside)   (melt)

The magnesite produced in the melt phase by this reaction may react further with orthopyroxene to produce olivine accompanied by decarbonation:

$$Mg_2Si_2O_6 + 2MgCO_3 = 2Mg_2SiO_4 + 2CO_2.$$
(enstatite)   (melt)   (ol)

We consider that such a process may be responsible for producing the metasomatic trend from lherzolite to olivine wehrlite seen, for example, in the suite of diopside-rich peridotite xenoliths in kimberlites (Waters and Erlank 1988). Significant modification of a primary carbonate melt may occur by interaction with a peridotite en route to the surface. For instance, decarbonation would consume the magnesitic component in the melt (second reaction above) to produce a more calcitic carbonatitic melt.

5 Conclusions

Our experiments permit the estimation of the compositions of melts produced by crystallizing carbonatite en route to the surface from depths of ca. 70 km.

1. The primary control over melt composition is degree of crystallization. The crystallizing assemblage is dominated by dolomite with lesser olivine and minor spinel for much of the interval between liquidus and solidus.

2. There is no evidence for liquid immiscibility in any of the experiments.

3. A close match to the Oldoinyo Lengai composition cannot be produced from the Wallace and Green (1988) "primary" carbonatite composition by crystal fractionation at shallower levels. Calculations suggest that any possible parent carbonatite to Oldoinyo Lengai must have more SiO_2, less FeO and a significantly higher K/Na ratio than CM1. We envisage that the choice of an appropriate source peridotite composition may change the equilibrium carbonatite melt composition significantly enough to make it possible to derive the Oldoinyo Lengai composition from a primary melt by closed-system processes.

4. Decarbonation of carbonatitic liquids at ca. 14 kb implies that a "primary" carbonatite composition cannot be preserved to the Earth's surface. This decarbonation reaction may provide a mechanism by which calcitic carbonatite magmas may be produced from primary (mantle-derived) carbonatites. Furthermore, open-system behaviour (e.g. the reaction of dolomitic melt with enstatite to produce diopside plus a more magnesitic melt) may alter the composition of any primary carbonate melt significantly. Decarbonation of the melt

at lower pressures also produces a fluid dominated by $H_2O + CO_2$ and containing significant Cl, Na and K, which may be the source of fenitizing fluids associated with many carbonatites.

Acknowledgements. We thank Simon Stephens for careful work preparing polished sections and Keith Harris and Wieslaw Jablonski for technical assistance. Peter Wyllie is thanked for helpful comments, and S. Foley and H. O'Neill for constructive reviews. The study was supported by the Australian Research Council and a South African FRD fellowship (RJS).

References

Baker MB, Wyllie PJ (1990) Liquid immiscibility in a nephelinite-carbonate system at 25 kb and implications for carbonatite origin. Nature 346:168–170

Cooper AF, Reid DL (1991) Textural evidence for calcite carbonatite magmas, Dicker Willem, south-west Namibia. Geology 19:1193–1196

Cooper AF, Gittins J, Tuttle OF (1975) The system Na_2CO_3-K_2CO_3-$CaCO_3$ at 1 kb and its significance in carbonatite petrogenesis. Am J Sci 275:534–560

Fanelli MF, Cava N, Wyllie PJ (1986) Calcite and dolomite without portlandite at a new eutectic in CaO-MgO-CO_2-H_2O, with applications to carbonatites. In: Morphology and phase equilibria of minerals. Proc 13th General Meeting Int Mineral Assoc Sofia, pp 313–322

Green DH, Wallace ME (1988) Mantle metasomatism by ephemeral carbonatite melts. Nature 336:459–462

Hogarth DD (1989) Pyrochlore, apatite and amphibole: distinctive minerals in carbonatite. In: Bell K (ed) Carbonatites – genesis and evolution. Unwin Hyman, London, pp 105–148

Irving AJ, Wyllie PJ (1975) Subsolidus and melting relationships for calcite, magnesite and the join $CaCO_3$-$MaCO_3$ to 36 kb. Geochim Cosmochim Acta 39:35–53

Jago BB, Gittins J (1991) The role of fluorine in carbonatite magma evolution. Nature 349:56–58

Keller J (1981) Carbonatite volcanism in the Kaiserstuhl alkaline complex: evidence for highly fluid carbonatitic melts at the earth's surface. J Volcanol Geotherm Res 9:423–431

Keller J (1989) Extrusive carbonatites and their significance. In: Bell K (ed) Carbonatites – genesis and evolution. Unwin Hyman, London, pp 70–88

Keller J, Krafft M (1990) Effusive natrocarbonatite activity of Oldoinyo Lengai, June 1988. Bull Volcanol 52:629–645

Kjarsgaard BA, Hamilton DL (1989) The genesis of carbonatites by immiscibility. In: Bell K (ed) Carbonatites – genesis and evolution. Unwin Hyman, London, pp 388–404

Reid DL, Cooper AF (1992) Oxygen and carbon isotope patterns in the Dicker Willem carbonatite complex, southern Namibia. Chem Geol 94:293–305

Taylor WR, Green DH (1987) The petrogenetic role of methane: effects on liquidus phase relations and the solubility of reduced C-H-O volatiles. In: Mysen BO (ed) Magmatic processes and physiochemical principles. Geochem Soc Spec Publ 1:121–138

Wallace ME, Green DH (1988) An experimental determination of primary carbonatite magma composition. Nature 335:343–346

Waters FG, Erlank AJ (1988) Assessment of the vertical extent and the distribution of mantle metasomatism below Kimberley, South Africa. J Petrol Spec Lithos Issue: 185–204

Wyllie PJ (1989) Origin of carbonatites: evidence from phase equilibrium studies. In: Bell K (ed) Carbonatites – genesis and evolution. Unwin Hyman, London, pp 500–545

Wyllie PJ, Huang W-L (1976) Carbonation and melting reactions in the system CaO-MgO-SiO_2-CO_2. Geology 3:621–624

Wyllie PJ, Huang W-L, Otto J, Byrnes AP (1983) Carbonation of peridotites and decarbonation of siliceous dolomites represented in the system CaO-MaO-SiO_2-CO_2 to 30 kbar. Tectonophysics 100:359–388

Subject Index

Springer-Verlag
and the Environment

We at Springer-Verlag firmly believe that an international science publisher has a special obligation to the environment, and our corporate policies consistently reflect this conviction.

We also expect our business partners – paper mills, printers, packaging manufacturers, etc. – to commit themselves to using environmentally friendly materials and production processes.

The paper in this book is made from low- or no-chlorine pulp and is acid free, in conformance with international standards for paper permanency.